L'Organisation de Coopération et de Développement Économiques (OCDE), qui a été instituée par une Convention signée le 14 décembre 1960, à Paris, a pour objectif de promouvoir des politiques visant :
— à réaliser la plus forte expansion possible de l'économie et de l'emploi et une progression du niveau de vie dans les pays Membres, tout en maintenant la stabilité financière, et contribuer ainsi au développement de l'économie mondiale;
— à contribuer à une saine expansion économique dans les pays Membres, ainsi que non membres, en voie de développement économique;
— à contribuer à l'expansion du commerce mondial sur une base multilatérale et non discriminatoire, conformément aux obligations internationales.

Les Membres de l'OCDE sont : la République Fédérale d'Allemagne, l'Australie, l'Autriche, la Belgique, le Canada, le Danemark, l'Espagne, les États-Unis, la Finlande, la France, la Grèce, l'Irlande, l'Islande, l'Italie, le Japon, le Luxembourg, la Norvège, la Nouvelle-Zélande, les Pays-Bas, le Portugal, le Royaume-Uni, la Suède, la Suisse et la Turquie.

L'Agence de l'OCDE pour l'Énergie Nucléaire (AEN) a été créée le 20 avril 1972, en remplacement de l'Agence Européenne pour l'Énergie Nucléaire de l'OCDE (ENEA) lors de l'adhésion du Japon à titre de Membre de plein exercice.

L'AEN groupe désormais tous les pays Membres européens de l'OCDE ainsi que l'Australie, le Canada, les États-Unis et le Japon. La Commission des Communautés Européennes participe à ses travaux.

L'AEN a pour principaux objectifs de promouvoir, entre les gouvernements qui en sont Membres, la coopération dans le domaine de la sécurité et de la réglementation nucléaires, ainsi que l'évaluation de la contribution de l'énergie nucléaire au progrès économique.

Pour atteindre ces objectifs, l'AEN :
— *encourage l'harmonisation des politiques et pratiques réglementaires dans le domaine nucléaire, en ce qui concerne notamment la sûreté des installations nucléaires, la protection de l'homme contre les radiations ionisantes et la préservation de l'environnement, la gestion des déchets radioactifs, ainsi que la responsabilité civile et les assurances en matière nucléaire;*
— *examine régulièrement les aspects économiques et techniques de la croissance de l'énergie nucléaire et du cycle du combustible nucléaire, et évalue la demande et les capacités disponibles pour les différentes phases du cycle du combustible nucléaire, ainsi que le rôle que l'énergie nucléaire jouera dans l'avenir pour satisfaire la demande énergétique totale;*
— *développe les échanges d'informations scientifiques et techniques concernant l'énergie nucléaire, notamment par l'intermédiaire de services communs;*
— *met sur pied des programmes internationaux de recherche et développement, ainsi que des activités organisées et gérées en commun par les pays de l'OCDE.*

Pour ces activités, ainsi que pour d'autres travaux connexes, l'AEN collabore étroitement avec l'Agence Internationale de l'Énergie Atomique de Vienne, avec laquelle elle a conclu un Accord de coopération, ainsi qu'avec d'autres organisations internationales opérant dans le domaine nucléaire.

FOREWORD

It is generally agreed that the most promising solution for the disposal of long-lived radioactive wastes is the emplacement into suitable, deep geological formations.

Isolation of radioactive materials is assured by at least three barriers :

- the waste form ;
- the waste container ;
- the disposal formation.

Safety analyses normally assume that over the long term ground water may come into contact with radioactive waste as a consequence of the failure of the various barriers. After ground water has reached the waste, leaching starts and migration of radionuclides follows.

The assessment of the possible consequences of this breach of containment requires a good understanding of the physico-chemical mechanisms that control the migration of radionuclides through the disposal formation and the geologic materials that intervene between the disposal formation and the surface.

Most radionuclides carried by ground water interact with the geologic materials and, as a result, their migration can be significantly slower than the ground water velocity. The magnitude of radionuclides retardation can vary within extremely large limits as a function of many factors.

Research programmes are being pursued in several countries to investigate the migration of radionuclides in geological formations. Recognizing that scientists working in this field need to meet periodically in order to discuss their respective experiences and co-ordinate future work, the Commission of the European Communities and the OECD Nuclear Energy Agency decided to organize this workshop. These proceedings represent a record of the papers and the discussions at the meeting.

Proceedings of the Workshop on

THE MIGRATION OF LONG-LIVED RADIONUCLIDES IN THE GEOSPHERE

Brussels, 29th-31st January, 1979

Compte rendu d'une réunion de travail sur

LA MIGRATION DES RADIONUCLÉIDES À VIE LONGUE DANS LA GÉOSPHÈRE

Bruxelles, 29-31 janvier 1979

jointly organised by the
OECD NUCLEAR ENERGY AGENCY
and the
COMMISSION OF THE EUROPEAN COMMUNITIES

organisé conjointement par
l'AGENCE DE L'OCDE POUR L'ÉNERGIE NUCLÉAIRE
et la
COMMISSION DES COMMUNAUTÉS EUROPÉENNES

ORGANISATION FOR ECONOMIC CO-OPERATION AND DEVELOPMENT
ORGANISATION DE COOPÉRATION ET DE DÉVELOPPEMENT ÉCONOMIQUES

The Organisation for Economic Co-operation and Development (OECD) was set up under a Convention signed in Paris on 14th December, 1960, which provides that the OECD shall promote policies designed:

— to achieve the highest sustainable economic growth and employment and a rising standard of living in Member countries, while maintaining financial stability, and thus to contribute to the development of the world economy;

— to contribute to sound economic expansion in Member as well as non-member countries in the process of economic development;

— to contribute to the expansion of world trade on a multilateral, non-discriminatory basis in accordance with international obligations.

The Members of OECD are Australia, Austria, Belgium, Canada, Denmark, Finland, France, the Federal Republic of Germany, Greece, Iceland, Ireland, Italy, Japan, Luxembourg, the Netherlands, New Zealand, Norway, Portugal, Spain, Sweden, Switzerland, Turkey, the United Kingdom and the United States.

The OECD Nuclear Energy Agency (NEA) was established on 20th April 1972, replacing OECD's European Nuclear Energy Agency (ENEA) on the adhesion of Japan as a full Member.

NEA now groups all the European Member countries of OECD and Australia, Canada, Japan, and the United States. The Commission of the European Communities takes part in the work of the Agency.

The primary objectives of NEA are to promote co-operation between its Member governments on the safety and regulatory aspects of nuclear development, and on assessing the future role of nuclear energy as a contributor to economic progress.

This is achieved by:

— *encouraging harmonisation of governments' regulatory policies and practices in the nuclear field, with particular reference to the safety of nuclear installations, protection of man against ionising radiation and preservation of the environment, radioactive waste management, and nuclear third party liability and insurance;*

— *keeping under review the technical and economic characteristics of nuclear power growth and of the nuclear fuel cycle, and assessing demand and supply for the different phases of the nuclear fuel cycle and the potential future contribution of nuclear power to overall energy demand;*

— *developing exchanges of scientific and technical information on nuclear energy, particularly through participation in common services;*

— *setting up international research and development programmes and undertakings jointly organised and operated by OECD countries.*

In these and related tasks, NEA works in close collaboration with the International Atomic Energy Agency in Vienna, with which it has concluded a Co-operation Agreement, as well as with other international organisations in the nuclear field.

621.4838
WOR

AVANT-PROPOS

Il est reconnu d'une manière générale que la solution la plus prometteuse pour l'évacuation des déchets radioactifs à vie longue en est leur dépôt dans des formations géologiques profondes appropriées.

L'isolation des matières radioactives est assurée par au moins trois barrières :

- la forme des déchets ;
- le conteneur des déchets ;
- la formation destinée à l'évacuation des déchets.

Les analyses relatives à la sûreté supposent qu'à long terme les eaux souterraines peuvent entrer en contact avec les déchets radioactifs, suite à une défaillance des diverses barrières. Après que les eaux souterraines aient atteint les déchets, la lixiviation commence et la migration des radionucléides suit.

L'estimation des conséquences éventuelles de cette rupture du confinement nécessite une bonne compréhension des mécanismes physico-chimiques qui contrôlent la migration des radionucléides au travers de la formation géologique et des matériaux géologiques qui se trouvent entre la formation renfermant les déchets et la surface.

La plupart des radionucléides transportés par les eaux souterraines réagissent avec les matériaux géologiques et, en conséquence, leur migration peut être nettement plus lente que l'écoulement des eaux souterraines. L'importance du retard des radionucléides peut varier à l'intérieur de limites très larges et être fonction de plusieurs facteurs.

Des programmes de recherche sont en cours dans plusieurs pays pour étudier la migration des radionucléides dans les formations géologiques. Reconnaissant la nécessité qu'ont les chercheurs dans ce domaine de se rencontrer périodiquement afin de discuter leurs expériences respectives et de coordonner les travaux futurs, la Commission des Communautés Européennes et l'Agence de l'OCDE pour l'Energie Nucléaire ont décidé d'organiser cette Réunion de travail. Le présent compte rendu comprend les communications qui y ont été présentées ainsi que la totalité des discussions dont elles ont fait l'objet.

CONTENTS

TABLE DES MATIÈRES

Session III - PLUTONIUM AND ENVIRONMENT

Séance III - PLUTONIUM ET ENVIRONNEMENT

Chairman - Président : Dr. D. RAI (United States)

Session IV - TRANSPORT IN VARIOUS MEDIA

Séance IV   - TRANSPORT DANS DIFFERENTS MILIEUX

Chairman - Président : Dr. F. GIRARDI (CEC)

Session V   - <u>ION EXCHANGE AND MIGRATION</u>

Séance V   - <u>ECHANGE D'IONS ET MIGRATION</u>

Chairman - Président : Dr. F.P. SARGENT (Canada)

Session I
Models — Radionuclides Transport

Chairman — Président
M. R.H. HEREMANS
(Belgique)

Séance I
Modèles — Transport des radionucléides

A GEOHYDROLOGICAL MODEL FOR THE LONG-TERM RISK ANALYSIS
OF THE DISPOSAL OF RADIOACTIVE WASTE IN SALT DOMES IN
THE NETHERLANDS

P. Glasbergen
National Institute for Water Supply
P.O. Box 150
2260 AD   Leidschendam
Netherlands

ABSTRACT

The geohydrological situation in the north-eastern part of the
Netherlands is influenced by the presence of a number of salt domes.
The permeability and the chemical characteristics of the aquifers are
dependent on the depth. As a subject of this study a salt dome is
chosen, which is representative.
Processes leading to a failure of the rock-salt barrier can be the
dissolution of salt (subrosion) and the upward movement of the salt
dome (diapirism). With the aid of a computer model the dissolution
rate is calculated.
Some input data are determined with steady state ground-water flow
models. The effect of climatic changes, land subsidence and variation
of sea-level is taken into account.
Calculations are carried out to investigate the sensitivity of the
dissolution-model for variations in input-data. The combination of
failure modes is described. The worst case and the most probable case
for the release mode of radionuclides is indicated.

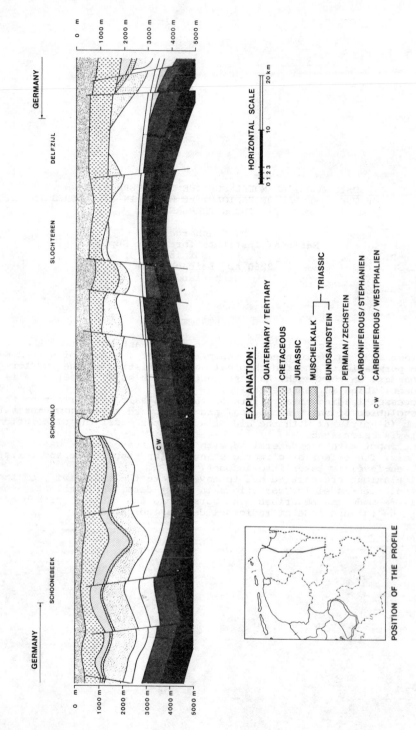

figure 1
geological profile over N.E. Netherlands

1.   INTRODUCTION

For the disposal of radioactive waste in the Netherlands salt domes occurring at attainable depths seem to be promising. Other geological formations such as: clay and limestone are considered unsuitable.

The aim of this study is to investigate the suitability of a salt dome as a disposal site with respect to its stability in geohydrological and geological point of view.

In this study a generic model is presented to investigate the effectiveness of the rock salt containment as a barrier. On basis of a preliminary study of the Geological Survey of the Netherlands a system of models is developed to analyse the ground-water flow and its consequences for the salt dome. With help of the distinct models the process of subrosion of a salt dome in combination with natural geological processes is analysed.
The model-system results in a general view of the release modes.

Based on the model calculations of this study the Institute for Atomic Science in Agriculture has worked out the pathways through the biosphere and the risks for future generations of man.

2.   DESCRIPTION OF THE GEOHYDROLOGICAL SITUATION IN THE NORTH-EASTERN PART OF THE NETHERLANDS AS INFLUENCED BY THE SALT STRUCTURES

A simplified geological profile of the north-eastern part of the Netherlands is presented in figure 1. The salt-layers are found at a depth of 2000 to 3000 m below the surface. Locally the salt-layers are transformed into salt pillows and salt domes. The salt domes have, in particular, influenced the structure of the overlying formations. Depressions can often be observed around a salt dome due to the migration of the underlying salt into the dome.

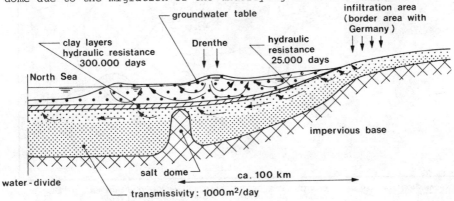

Explanation:

Holocene, Pleistocene, Pliocene : sand (and clay)	good - moderate	permeable
Miocene, Oligocene : clay (and sand)	bad	"
Eocene, Paleocene : sand (and clay)	moderate	"
Mesozoic : chalk, clay, sand, limestone	bad	"
Zechstein : rock salt, anhydrite	impervious	

figure 2
schematic geohydrological cross-section of the N.E. Netherlands (NW-SE)

From a geohydrological point of view all deposits above the salt structures have to be considered as water containing strata. However the porosity of the sandy layers, generally speaking, decreases with depth due to compression and chemical deposition in the pores. Also the clay layers at greater depths are compressed in such a way, that they are almost impermeable. The relatively good permeable aquifers are found in the coarse sands from the Tertiary and the Quaternary. As a consequence of the good permeability of these strata the greater part of ground-water flow takes place in the mentioned layers. However a small part of the ground water movement will appear in the deep aquifers.

Figure 2 gives an impression of the hydrologic cycle for the north-eastern part of the Netherlands. The border region with the Federal Republic of Germany and the topographic high parts of the north-east can be considered as infiltration areas.

Locally the deeper aquifers are interrupted by the salt domes. Only a flow around these salt structures would be possible. Also the good permeable aquifers are influenced by the salt structures only relatively thin layers of Quaternary and Tertiary deposits are found above the salt domes. Figure 3 presents the contour lines of the base of the good permeable strata in the coastal area of the Netherlands.

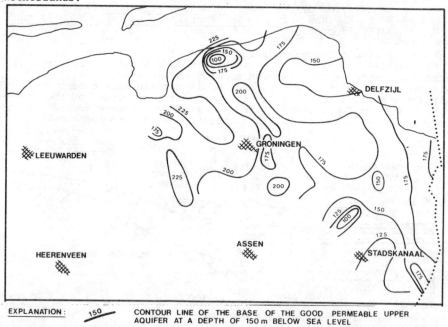

EXPLANATION: 150 ——— CONTOUR LINE OF THE BASE OF THE GOOD PERMEABLE UPPER AQUIFER AT A DEPTH OF 150 m BELOW SEA LEVEL

0   10km
SCALE : ▬▬▬▬▬

figure 3
base of the good permeable upper aquifer

The chemical characteristics of the ground water depend on the depth of the aquifer. The aquifers of good permeability above the base (found in figure 3) contain fresh water. Only in a small zone along the coast is salt water found.
Below the base the aquifers contain salt water with chloride contents increasing with depth from a few hundred milligrams per litre till a few grams per litre.
The increasing salt concentration is coupled with a higher specific density.

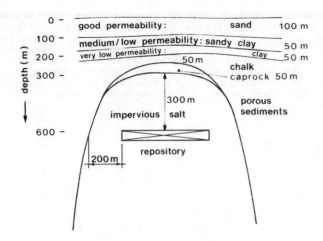

figure 4
cross-section over the model salt dome and adjacent strata

For a more elaborate geohydrological modelling of the processes which involve ground-water movement a salt dome is chosen as
a subject of study as presented in figure 4. The data used are comparable with those for the salt domes which are suitable for the storage of radioactive waste. The top of the salt is situated at a depth
of 300 m, covered with 50 m caprock. Above the caprock strata are
found which are driven up due to diapirism. Between 150 and 200 m a
clay-layer is present, which separates the fresh water aquifer from
the deeper aquifers which contain salt water.
The radioactive waste (solidified into glass) is stored at a depth of
600 to 680 m.

3.   NATURAL SLOW PROCESSES LEADING TO A FAILURE OF THE SALT
      BARRIER

Considering the geology of the salt structures and the surrounding strata, different phenomena can be observed. The salt domes
are the result of an upward movement so it is necessary to investigate what can be expected in the future. Also it is obvious that water
in contact with salt will lead to dissolution. The process of sub-
surface dissolution of rock salt is called subrosion. Besides these
phenomena which are a result of the presence of the salt structures,
tectonic movements and changes in the hydrologic cycle due to climatic changes are also involved.

The migration of salt into pillows or diapirs and the upward movement of salt can occur after a certain thickness of overlying soil material is reached. For the diapirism of salt domes it is
easier to give the average velocity over longer periods of, for instance, a million year than for shorter periods. Looking at shorter
periods we have to take into account the pulsating effect of the
process. The Geological Survey of the Netherlands has calculated that
the average velocity of diapirism during the last 66 million years
(Tertiary) is 0.05 mm/a. Due to pulsating effects a velocity of 0.25
mm/a might be possible.
As a maximum figure, known from literature for salt domes elsewhere
in the world,an upward velocity of 2.5 mm/a has been taken into
account [1].

The deposition of sand and clay most often happens in rivers
and coastal areas. Most parts of the Netherlands lie in a large area
with tectonic subsidence having taken place during the last millions
of years. It is to be expected that the tectonic subsidence will

continue for the forthcoming periods. Extrapolation for the next
250,000 years leads to the picture described in figure 5. The last
thousand years the lower parts of the Netherlands are protected
against flooding by dikes. However the heightening of the dikes will
be limited. If the sea-level is constant it will be clear that the
salt dome region is flooded. There is only compensation possible if
the opportunity of sedimentation in a sea exists. So in the long run,
a coastal landscape just below or just above sea-level can be expected.

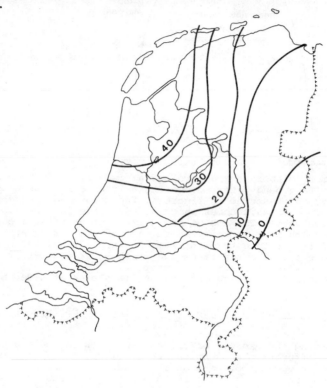

EXPLANATION:

30   SUBSIDENCE OF 30 METER

SCALE: 10 km

figure 5
predicted subsidence after 250,000 years extrapolated from beginning Quaternary

Looking at the global climate in the past one sees an undu-
lation in relatively short periods of 10^4 - 10^5 years. From geological
investigations in the Netherlands the present climate can be considered as an average one [2]. It might change in the future into either
a relatively little colder or little warmer climate. A warmer climate
will result in a higher sea-level for the Netherlands, so that the
effect of land-subsidence will be intensified.

4.   MODELLING GROUND-WATER FLOW AND THE DISOLUTION OF SALT

4.1. Summary of models

The aim of the use of geohydrological models is to provide the instruments to calculate the transport time of radionuclides to the biosphere and the moment of release of nuclides from the barrier of the rock salt containment. For the different processes in the pathways from the repository to the biosphere and for the failure modes of the barriers a set of models is prepared.

Figure 6 gives an indication of the system of models used. The first four mentioned models are discussed in this paper. The last group of models is studied by the Institute for Atomic Sciences in Agriculture (ITAL) [3].

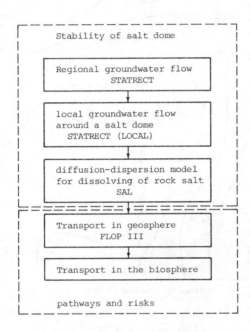

figure 6
models used in the different phases of calculation

4.2. Models for regional ground-water flow

Starting with the last mentioned problem model techniques have been developed by the National Institute for Water Supply to determine three dimensional ground-water flow. The model used is based on the method of finite elements. To make an approximate calculation for the flow velocity in the aquifers around the salt dome the following simplifications are applied.
The aquifer system is simplified into a two layer system i.e. a relatively good permeable upper layer and a relatively bad permeable deep layer. The permeability of the upper aquifer is well known from a number of pumping tests [4]. Boundary conditions for the flow model are:

- the phreatic ground-water levels
- the amount of water infiltrating in the border area with the
  Federal Republic of Germany
- the level of the North Sea

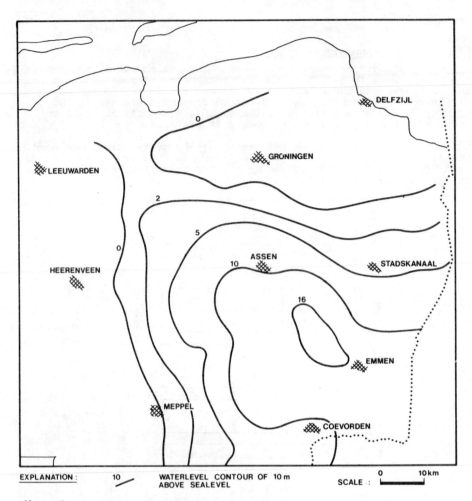

EXPLANATION :   10 ╱   WATERLEVEL CONTOUR OF 10 m
                                 ABOVE SEALEVEL   SCALE : 0   10 km

figure 7
ground-water contour lines of the upper aquifer

So with the help of the above mentioned data it is possible
to calculate ground-water flow for the deep aquifer. In figure 7
the contour-lines of the ground-water level in the upper aquifer are
presented based on field determinations. Figure 8 shows the calculated contour-lines for the deep aquifer, taking in account the land-
subsidence. From the displacement of the contour-lines it becomes
evident that with a lower level of the surface the ground-water flow
decreases. In the present situation the average velocity for the flow
in the deep aquifer (from 200 to 600 m) is about 0.1 m/a. It should
be pointed out that this is an average velocity. Due to inhomogeneity
of the strata a certain deviation is possible.

4.3. <u>Models for local ground-water flow.</u>

In the surroundings of a salt dome the horizontal flow lines
will bend around the diapir (figure 9). Due to these curved flow
lines the flow velocity can vary between 0 and 0.2 m/a at a regional
flow velocity of 0.1 m/a.

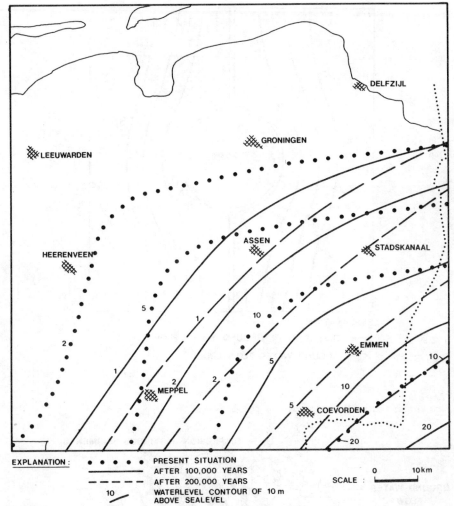

EXPLANATION :
- • • • • • PRESENT SITUATION
- — — — — AFTER 100,000 YEARS
- – – – – – AFTER 200,000 YEARS
- 10 —— WATERLEVEL CONTOUR OF 10 m ABOVE SEALEVEL

SCALE : 0 ——— 10km

figure 8
calculated contour lines of the deep aquifer

4.4. The model for salt solution

To describe the process of the dissolving of rock salt the model of figure 10 is used. The rock salt will probably be covered by a residue layer of insoluble substances. Estimations show that rock salt may contain 10 % of insoluble substance. The residue layer itself is impermeable. However there might be microfractures due to movements in the rock salt. The thickness of the layer is estimated at the flanks on 1 m as a result of previous dissolution. At the top of the diapir a layer of 50 m caprock covers the rock salt. Micro-fractures may also occur in this caprock. These microfractures might be caused by an expansion of the salt dome due to the release of the decay heat from the high-level reprocessing waste.

The process of solution is governed by the diffusion equation for the salt transport through the microfractures, where the fluid is in rest and by the dispersion equation in the porous sediments outside the covering boundery layer.
The equations are:

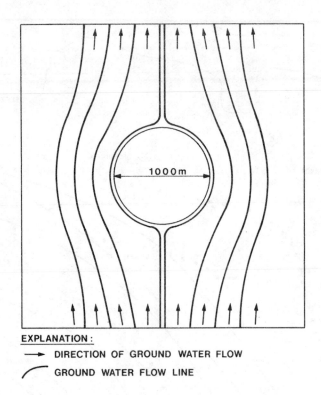

EXPLANATION:

→ DIRECTION OF GROUND WATER FLOW

⌒ GROUND WATER FLOW LINE

figure 9
flow pattern around a salt dome in the deep aquifer

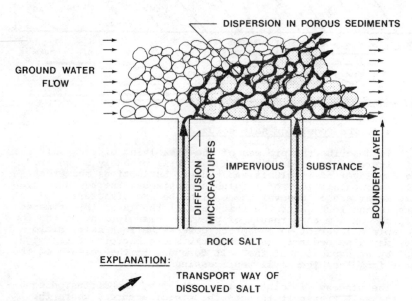

EXPLANATION:

➚ TRANSPORT WAY OF
DISSOLVED SALT

figure 10
model for dissolving of salt by diffusion and dispersion

diffusion : $\frac{\delta c}{\delta t} = D_{diff} \frac{\delta^2 c}{\delta y^2}$

and dispersion:

$$\frac{\delta c}{\delta t} + v \frac{\delta c}{\delta x} - D_{disp} (\frac{\delta^2 c}{\delta x^2} + \frac{\delta^2 c}{\delta y^2}) = 0$$

where c = concentration of salt in water
      v = velocity of water flow in the pores
      t = time

For the diffusion constant D_{diff} a value of 0.025 m^2/a in porous sediment can be taken, while the dispersion constant D_{disp} is related to the flow velocity and the inhomogeneities in the porous medium. For instance at a flow velocity of 0.2 m/a the D_{disp} can vary in between 0.02 and 0.2 m^2/a.

The model calculates in a rectangular grid the transport of salt by ground-water under steady flow conditions. At the flanks the grid is placed horizontal and at the top of the dome vertical. At one boundary of the grid the concentration of salt is coupled to the diffusion in the boundary layer. The model calculates for the grid the concentrations dissolved salt and adds up the discreet concentrations to get the total dissolved salt transport downstream. Figure 11 shows the different concentration lines.

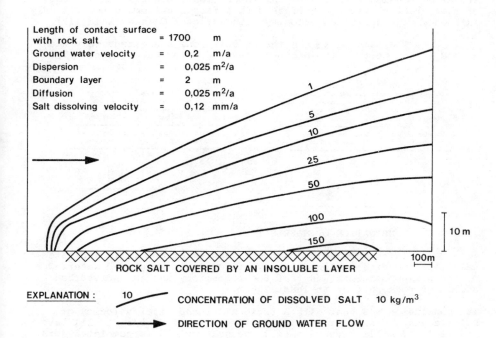

Length of contact surface with rock salt = 1700 m
Ground water velocity = 0,2 m/a
Dispersion = 0,025 m^2/a
Boundary layer = 2 m
Diffusion = 0,025 m^2/a
Salt dissolving velocity = 0,12 mm/a

ROCK SALT COVERED BY AN INSOLUBLE LAYER

EXPLANATION : 10 — CONCENTRATION OF DISSOLVED SALT 10 kg/m^3
— DIRECTION OF GROUND WATER FLOW

figure 11
concentration of dissolved salt in a horizontal plan adjacent to the salt dome

The results lead to the values for salt transport summarised in table I. With a flow velocity of 0.2 m/a the subrosion velocity amounts 0.12-0.16 mm/a with a variation of dispersion constants from 0.025 till 0.2 m^2/a.

Table I
Calculated rate of dissolution of a salt dome

	Situation	flow velocity (m/a)	diffusion factor (m²/a)	boundery layer (m)	dispersion factor (m²/a)	rate of dissolution (mm/a)
top of the dome	Slow flowing groundwater	0,5	0,025	50	0,05 – 0,5	0,008
flanks of the dome	Slow flowing groundwater	0,2	0,025	2	0,025– 0,2	0,12-0,16
	moderate/fast flowing ground-water	2 20	0,025 0,025	2 2	0,2 – 2 2 –20	0,18-0,19 0,19

The rate of dissolution of salt at the top of the dome through the caprock is only a fraction of the dissolution rate at the flank (in order of 0.008 mm/a).

The sensitivity of the models for calculation the salt dissolution was investigated by varying the different input data. Important data are:

- the dispersion constant as mentioned before
- the thickness of the boundary layer
- the relative surface of the microfractures in the boundary layer

The results show that the subrosion velocity is almost linearly proportional decreasing with rising thickness (see table II). An increase of the relative fracture surface with a factor 5 from 10 to 50 % only influences the subrosion velocity with a factor 3 (see table III).

The assumptions and the calculations are extensively described in a report of the National Institute for Water Supply [5].

Table II

The effect of the thickness of the boundery layer at the flank on the dissolution rate.
Groundwater flow velocity = 20 m/a
Dispersion factor   = 2 m²/a

boundery layer (m)	dissolution rate of rock salt (mm/a)
0,2	1,83
0,5	0,76
1,0	0,38
2,0	0,19
5,0	0,08
10	0,04

Table III

The effect of the relative fracture surface on the dissolution rate.
Groundwater flow velocity = 0.2 m/a
Dispersion factor   = 0.2 m²/a

relative fracture surface (%)	dissolution rate of rock salt (mm/a)
1	0,02
10	0,16
50	0,48
70	0,55

5.   COMBINATION OF MODELS

In the preceding chapter a system of models is indicated to calculate the dissolution velocity of rock salt by ground water. One of the presuppositions was that the repository has an undisturbed shield of rock salt of at least 200 m.

The conclusion was that with a regional ground-water velocity of 0.1 m/a and a dissolution velocity of 0.12 - 0.16 mm/a it will take 1.2 till 1.7 million years untill the salt shield is completely dissolved.

As stated before we have to take into account the effects on the hydrological system of slow geological processes and changes in climate. A warmer climate will result in a rise in sea-level and a decreasing ground-water flow in the salt dome area. The dissolution velocity will also decrease but even in ground water at rest diffusion occurs. Looking at the already mentioned very long times it makes no sence to work out this situation.

A colder climate could lower the sea-level and enlarge the ground-water flow a little. However, even a regional flow velocity of 1 m/a leads to a dissolution velocity of 0.18 mm/a. So the alteration in flow has relatively less influence on the salt containment shield.

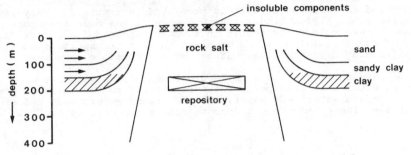

EXPLANATION:

figure 12   →   **DIRECTION OF GROUND WATER FLOW**
situation after an upward movement of 450 m

The slow geological processes can have an important influence on the salt structure.
Of great importance is the diapiric movement leading to bulk transport of the waste together with the surrounding salt. A rapid rising of the salt dome could lead to contact with better permeable aquifers (figure 13). In figure 13 a diagram is presented of the time-depending changes of the position of the repository. The most probable situation is the left branch of the diagram. However, as long as there is no exact figure for the velocity of diapirism it is necessary also to take into account the high velocity leading to the situation at the right side.

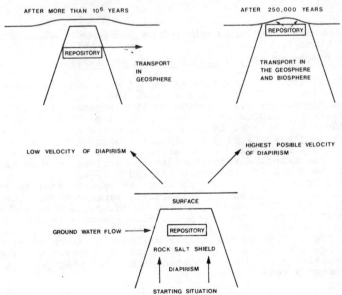

figure 13
time-depending changes of the position of the repository

Calculations show that combinations of rising at a velocity of 0.25-0.5 mm/a and subrosion lead to a total dissolution of the salt shield after at least 1 million years.

An evaluation of the different combinations of diapiric velocity and dissolution of salt is presented in figure 14. Only at velocities of diapirism higher than 0.5 mm/a is the bulk transport dominant. Lower velocities result in a break-through of the rock salt shield of the disposal mine. After break-through transport of nuclides leaching from the glass-matrix becomes possible.

The worst case is a rapid diapirism with a velocity of 2.5 mm/a. After bulk transport of the repository to the surface spreading in surface water, sea water, shallow ground-water, soil and air is possible. These biosphere transport models are studied by ITAL [3].

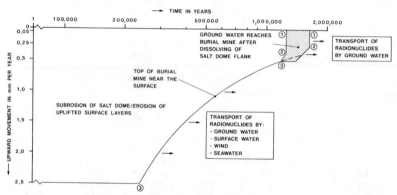

figure 14
dependency of release time of buried radionuclides from velocity of upward movement of salt dome

6.   TRANSPORT OF RADIONUCLIDES IN THE GEOSPHERE

Describing the geosphere transport knowledge of water quality characteristics is of utmost importance. The ground-water that enters in the repository will be completely saturated with salt. The specific density of this water is about 1250 kg/m^3. A high density influences the flow pattern. It is almost impossible for saturated water to move to the surface. Only diffusion through the overlying strata to the surface remains as a natural transport mechanism. The vertical distance from the repository to the surface is then taking into account the diapiric movement still 300 till 425 m.

As a human activity ground-water extraction near to the repository can not be excluded. Contaminated water can only occur, due to the high specific density, at the depth of the disposal mine. To pump up contaminated water a pumping filter with a length of more then 300 m is necessary. The result might be an intrusion of salt water in the lower parts of the filter mixing with fresh water out of the upper aquifer. The spreading of contaminated salt water coming out of the repository can be calculated with a computer program that follows different stream lines.

7.   LITERATURE

(1)   Geological Survey of the Netherlands.
      Report: Enkele geologische gegevens voor het opstellen van een
      voorlopig model ten dienste van de risico-analyse van het opbergen
      van radioactief afval in een zoutkoepel in N.O.-Nederland.
      Haarlem, 1977.

(2)   Geological Survey of the Netherlands:
      Toelichting bij geologische overzichtskaarten van Nederland.
      Haarlem, 1975.

(3)   Dorp, F. van, et al:
      Transport of radionuclides stored in a salt dome to and through
      the biosphere.
      Institute for Atomic Science in Agriculture.
      Wageningen, 1978.

(4)   Werkgroep Regionaal Geohydrologisch Onderzoek in de Provincie
      Drenthe:
      Regionaal Geohydrologisch Onderzoek in de Provincie Drenthe.
      Voorburg, 1978.

(5)   National Institute for Water Supply:
      Een voorlopig geohydrologisch model ten behoeve van de veilig-
      heidsanalyse van de opslag van radioactieve afvalstoffen in een
      zoutkoepel.
      Report: g.h.a.-r-77/04.
      Voorburg, 1977.

Discussion

L.R. DOLE, United States

Does your model reconcile the quantity and composition of a particular cap rock with the composition and estimated dissolution of a particular salt diapir ?

P. GLASBERGEN, The Netherlands

The content of insoluble substances in rock salt in The Netherlands is about 10 %. The cap rock is a result of previous dissolution of salt during millions of years. The thickness of cap rock is correlated with the diapiric movement and the salt dissolution rate.

L.R. DOLE, United States

In regard to the risk analysis, have you studied the expected distributions of growth rate of the diapirs, the permeabilities, the dissolution rates, the subsidence rates, etc., in order to determine the probabilities of the envisaged events ?

P. GLASBERGEN, The Netherlands

I think that we can deal with some processes in a more or less probabilistic way. But subrosion for instance is a deterministic process. I will not exclude the possibility that in the future we get better data based on field experiments and drillings, which will give us the possibility to calculate several geological and hydrological processes in a deterministic way.

G. MATTHESS, Federal Republic of Germany

Do you think that the hazard of the repository will change after flooding by the sea ?

P. GLASBERGEN, The Netherlands

In case of a closed repository the hazard evaluation is not changed. Only during the operational period flooding by the sea can lead to water entering the mine. The result would be that the mine is filled with a saturated brine with high specific density. However the chance of flooding by the sea during the operational period is extremely low.

E. BÜTOW, Federal Republic of Germany

What are the dimensions of the finite elements groundwater transport model and the grid block size ?

P. GLASBERGEN, The Netherlands

The finite element model consist of a rectangular grid. For the specific data please refer to reference n° 5 in the paper.

<u>V.E. Della LOGGIA</u>, CEC

        Do you use in your calculations coefficients taken from
the literature or you measure them directly in situ ?

<u>P. GLASBERGEN</u>, The Netherlands

        In the salt dissolution model one of the most difficult
parameters is the dispersion coefficient. It is a function of flow
velocity and the inhomogeneities in the aquifer. Until now we have
some values for dispersion down to a depth of approximately 200 m.
For greater depths the same range of values is used.

        The geohydrological parameters of the upper aquifer are
based on field determinations (pumping tests,etc.) and for the deep
aquifer on literature studies. However the regional ground water
flow models provide a check on the data used. The model presented
is a generic one. When the drilling program is carried out a specific model will be developed based on field determinations of the
different parameters.

<u>F. GERA</u>, Nuclear Energy Agency

        I would like to ask if there are plans in The Netherlands
to measure growth rates of salt diapirs and what technique will be
used ?

<u>P. GLASBERGEN</u>, The Netherlands

        We have planned a drilling program for geohydrological
investigations. After approval by the government 6 boreholes will be
drilled. Three of them to a depth of 200 m above the salt dome and
three around the dome to a depth of 500 m. Comparison of the geological formations found above and around the salt dome could lead
to a figure for the rate of diapirism and the variation in the rate
during the geological history.

<u>E. LEDOUX</u>, France

        Le phénomène de diffusion du sel à travers la couche limite
d'éléments insolubles me paraît être le facteur dominant de la mise
en solution du sel. Quels sont les moyens dont on dispose pour évaluer le coefficient de diffusion de la couche intermédiaire ?

<u>P. GLASBERGEN</u>, The Netherlands

        In The Netherlands field data exist on the diffusion of
dissolved salt from deep brackish ground waters towards shallow
fresh aquifers. The limiting factor is not the value of the diffusion
coefficient but the thickness of the layer of insoluble residues.
For this thickness a sensitivity analysis is worked out. Earlier
investigations of some salt domes show a layer of insoluble residues
on the flanks of the domes.

<u>P.C. LEVEQUE</u>, France

        L'effet de subsidence que vous avez évoqué provoquera une
surcharge de quelques MPa sur la zone du dépôt.

        Les phénomènes de consolidation, au sens géotechnique,
c'est-à-dire d'expulsion d'eau des couches surchargées, auront pour
effet l'entraînement de radionucléides avec cette eau expulsée.

Concevez-vous le résultat de ce phénomène de migration comme défavorable ou sans importance, à la suite de la submersion de la zone, par la mer ?

P. GLASBERGEN, The Netherlands

The greater part of The Netherlands lies in an area which has undergone tectonic subsidence for millions of years. It is my opinion that an additional subsidence of 10-20 m in the future will have a very small influence on the consolidation of the sandy sediments.

TRANSPORT OF RADIONUCLIDES STORED IN A SALT DOME TO AND THROUGH THE BIOSPHERE

F. van Dorp, M.J. Frissel, P. Poelstra
Foundation Institute for Atomic Sciences in Agriculture
P.O. Box 48, 6700 AA Wageningen, The Netherlands.

ABSTRACT

A worst case analysis of risks for future generations of man due to the storage of highly radioactive waste embedded in glass from 10^6 MWe·year nuclear power production in a salt dome in the north-east of The Netherlands is described.

The most probable geologic containment failure mechanisms are: 1) the dissolution of the salt around the waste and the dissolution of the waste in deep groundwater; 2) diapyrism of the salt dome.

A high rate of diapyrism and other natural processes like climatic changes, erosion and subsidence of the land are included in the formulation of critical scenarios and pathways, as described in a drinking water model, an inhalation model, an agricultural model, an external radiation model, and a marine model.

INTRODUCTION

The generation of electric power by nuclear fission leads to the production of highly radioactive waste. It is proposed to dispose of this waste, after solidification, into deep geologic formations. For this purpose, salt domes in the north-east of The Netherlands seem to be promising.

The aim of this study is a worst case analysis of the hazards for future generations of man due to the storage of highly radioactive waste, solidified into glass from 10^6 Mwe·year nuclear power production in a salt dome. The study is based on a general description of salt domes in the north-east of The Netherlands by the Geologic Survey. The National Institute for Water Supply established a geohydrological model of the surroundings of the salt dome to evaluate the process of subrosion of the salt dome at different rates of diapyrism.

DISPOSAL CONCEPT[1]

Fig. 1 shows the general features of a salt dome and the placing of the radioactive waste. The supposition is that 50 000 containers of 50 litres of glass are stored at a depth of 600 to 680 m in an area of 600 x 800 m². Each container of 50 litres (from 0.6 ton fuel charged to a reactor) is surrounded by 750 m³ rock salt. At least 200 m of rock salt is left undisturbed on all sides of the storage area.

GEOLOGIC CONTAINMENT FAILURE MECHANISMS

The most probable geologic processes which may cause a dispersion of the waste in the biosphere are: 1) geohydrologic processes by which the rock salt around the waste is dissolved in groundwater (subrosion) and 2) the rising of the salt dome under the pressure of surrounding sedimentary layers (diapyrism). Other processes or events like vulcanic activity, seismic activity, faulting and meteorite impact are either improbable for the north-east of The Netherlands in the considered time period, or have direct effects on man which are more drastic than a possible dispersion of the waste (meteorites). The influence of climatic changes, erosion and subsidence on the way of exposure will be considered in the next paragraph.

The National Institute for Water Supply calculated the rates of dissolution of the salt dome in deep groundwater to be in the range of 0.01 to 0.2 mm·a^{-1}. When no caprock is present at the top of the salt dome the dissolution rate in fast flowing superficial groundwater can amount to 2 mm·a^{-1}.

The rate of diapyrism of the salt dome can be about 0.05 mm·a^{-1} as an average over millions of years and 0.25 mm·a^{-1} during shorter periods. In literature some rates of diapyrism as high as 2.5 mm·a^{-1} are mentioned.

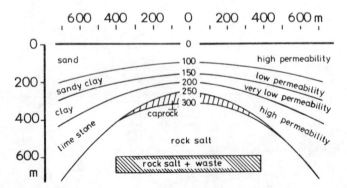

FIG. 1. Schematic presentation of a salt dome and the position of the waste.

---

[1] Calculations of the thermal effects of the waste led to lower loading densities than described in this study. These changes will be incorporated in a final site specific study.

Dissolution of the salt in deep groundwater can bring the waste in contact with this groundwater but this will take at least one million years. For rates of diapyrism which can be expected (up to 0.25 mm·a^{-1}) dissolution is the most probable containment failure mechanism. For rates of diapyrism higher than 0.5 mm·a^{-1}, however, dissolution of rocksalt by the water will be too slow to prevent the salt from rising to the surface. Because the aim of this study is a worst case analysis, it is based on the improbably high rate of diapyrism of 2.5 mm·a^{-1}, in which case the radioactive waste will reach the surface in about 250 000 years. As soon as the salt reaches the surface, dissolution by precipitation will become more important than dissolution in groundwater.

NATURAL PROCESSES LEADING TO DIFFERENT WAYS OF EXPOSURE

When the radioactive waste reaches the surface, one of the most important factors which determine the rate of release and the pathways through the biosphere, is the climate. Climate determines the character of the biosphere, and has strong influences on soil formation, erosion, geohydrological processes, changes in sealevel etc. Changes in climate for the future 250 000 to one million years can only be estimated from the past.

On the left hand side of fig. 2 climatic changes in the second half of the Pleistocene are presented. On the right hand side the position of the waste is indicated as a function of time for rates of diapyrism of 2.5 mm·a^{-1} (a) and 0.25 mm·a^{-1} (b). Line c indicates the position of the sealevel relative to the present-day ground level, when the rate of subsidence of the north-east of The Netherlands proceeds with 0.1 mm·a^{-1}. Till man constructed dykes in Holland this subsidence was in general compensated by sedimentation. As the maintenance of dykes in the future is not certain, the possibility that the north-east of The Netherlands remains land by sedimentation, as well as that it will be covered by sea has to be taken into account.

Depending on the amount of precipitation and on the permeability of the layers overlying the salt dome, the rising salt (worst case: 2.5 mm·a^{-1}) can dissolve somewhat below the surface, at the surface, or above the surface. In the latter case a hill will be formed.

All combinations of these processes can lead to different pathways for the radionuclides from the waste to man. In the next paragraph some critical combinations will be described.

CRITICAL SCENARIOS AND PATHWAYS

Radionuclides from the waste can cause a rise in the radiation dose to man when the waste dissolves in groundwater which is used as drinking water. As soon as the waste has reached soil surface, small fragments could be suspended in the air, and cause a rise in radiation doses when inhaled. Plant roots can absorb radionuclides from waste in the soil, introducing them to man through the foodchain. Radionuclides in the soil also cause a rise in radiation dose to man by external radiation. If the north-east of The Netherlands is submerged in the sea the radionuclides from the waste can also reach man via marine food products.

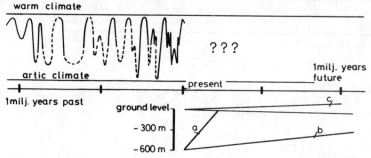

FIG. 2. Climatic changes in the past, position of the waste in future assuming a 2.5 mm·a^{-1} rise of the salt (a) or a 0.25 mm·a^{-1} rise (b) and the future sealevel (c) relative to the present day ground level.

For all these different pathways models are constructed, to calculate maximum dose rates to man. An estimate is made for each pathway of the number of exposed people. The calculations and assumptions are extensively described by Van Dorp et al. (1978).

Drinking water model

Two alternative hypotheses are tested. Firstly the assumption that all waste dissolves at the same time as the surrounding salt and that the water in which waste and salt are dissolved is diluted to drinking water. This results in drinking water with 25 to 76 mg glass per m^3 water if this water contains 200 to 600 g Cl per m^3. The second, more realistic, assumption is that the salt dissolves first, leaving the more unsoluble anhydrite and gypsum with the waste as a residual layer. When water percolates through this layer, waste as well as gypsum will slowly be dissolved. Water with 500 g SO_4 per m^3 (WHO-recommendation = 400 g·m^{-3}) can contain 40 mg glass per m^3. If this water is not influenced by the huge amounts of rocksalt below the residual layer, it could be used as drinking water. In both hypotheses, however, the quality of the drinking water would be poor compared to the quality of the water in the surroundings of the salt dome.

According to the ICRP-recommendations (ICRP 2, 6 and 9) for individual members of the public, drinking 1.2 litres per day containing 92 mg glass per m^3 with 250 000 year old waste, would still be acceptable. The nuclides contributing most to the risk to man are indicated in table I, of which ^{226}Ra is the most important.

Inhalation model

While the climate has no influence on the drinking water model, as long as people can live and drink water in The Netherlands, for the inhalation of contaminated dust the climate is of utmost importance. When the waste in the form of glass has reached the soil surface, it has, at least partly, to be ground to dust with particles smaller than 6 μm to become a risk for inhalation. This could happen underneath a glacier. As can be seen from fig. 2 ice ages with glaciers are to be expected for The Netherlands during the next million of years. During periglacial dry conditions after an ice age glass particles and other dust can be blown into the air and cause a concentration of about 7 x 10^{-9} g glass per cm^3 air. According to the ICRP-recommendations for individual members of the public, inhalation of air with a concentration of 8 x 10^{-9} g 250 000 years old glass per m^3 would still be acceptable. The most important radionuclides are given in table I, of which ^{229}Th causes the highest dose.

TABLE I. Composition of the waste and relative hazard of the different radio-nuclides present in 250 000 years old waste when ingested with drinking water or inhaled via the air. The relative risk is presented as the percentage of the total amount of water, respectively air in which the waste has to be diluted not to exceed the ICRP-recommendations.

| | Activity of the waste | | Relative hazard | |
	Bq·kg^{-1}	Ci·g^{-1}	% water	% air
Fission products				
^{99}Tc	1.0 x 10^9	2.8 x 10^{-5}	< 1	< 1
^{129}I*	6.1 x 10^6	1.6 x 10^{-7}	2	< 1
4N + 1 series				
^{237}Np	5.7 x 10^7	1.5 x 10^{-6}	3	12
^{233}U	3.9 x 10^7	1.0 x 10^{-6}	1	< 1
^{229}Th	3.9 x 10^7	1.1 x 10^{-6}	18	85
^{225}Ra	3.9 x 10^7	1.1 x 10^{-6}	9	< 1
^{225}Ac	3.9 x 10^7	1.1 x 10^{-6}	1	< 1
4N x 2 series				
^{230}Th	4.3 x 10^6	1.2 x 10^{-7}	< 1	1
^{226}Ra	4.3 x 10^6	1.2 x 10^{-7}	58	< 1
^{210}Pb	4.3 x 10^6	1.2 x 10^{-7}	6	< 1

*^{129}I is supposed to be included in the waste.

Agricultural models

Two processes for the incorporation of waste in soil are possible.

Firstly, when the salt is dissolved, a soil is formed on the remaining layer of gypsum and anhydrite containing the waste. If the amount of rather unsoluble compounds in the salt is 10% the remaining layer will contain 0.07% waste, which is supposed to consist of glass spheres of 1 cm diameter. The soil formed on this gypsum and anhydrite will not be very suitable for agriculture and a production of only 50 tons (fresh weight) of grass per ha per year is supposed to be possible. In this case the consumption of 1 litre of milk per day can lead to a radiationdose caused by ^{229}Ra amounting to 89% of the ICRP-recommendations. Consumption of meat will lead to a smaller dosis (see table II).

Secondly, after dissolution of the salt, the remaining layers with the waste may be eroded and spread over the surrounding area. This may result in a sedimentary soil with 0.005% waste (particle size of the glass after erosion is supposed to be 1 mm). The soil will probably be of good agricultural quality; continuing dissolution of the salt dome, however, could cause salinity problems. Because it is impossible to predict the diet of a population over 250 000 years, hence an "indicator crop" is used which could lead to a high transfer of radionuclides from soil to man. A production of 40 tons (fresh weight) of potatoes, or 130 tons (fresh weight) of grass per ha per year is assumed. In this case a consumption of 250 kg potatoes per person per year will lead to radiation doses from ^{226}Ra of about the level of the ICRP-recommendations. Data for meat consumption and for other radionuclides are given in table II.

For the transfer from grass to cattle and from cattle to milk or cattle to meat, factors mentioned by Burkholder et al. (1975) are used. Uptake of radionuclides by plants from soil is calculated in a new approach (fig. 3 and Van Dorp et al. 1979). Dissolution of radionuclides from the glass is a function of the diameter of the glass particles. Dissolved nuclides will be partly adsorbed on soil solid material according to their Kd-values (= concentration adsorbed/concentration in soil solution). The radionuclides in the soil solution can be

TABLE II. Radiation dose calculated in the different models, expressed as percentage of the ICRP-recommendations for individual members of the public. The assumed consumption per person per year is 365 kg milk, 100 kg meat, 250 kg potatoes, 36.5 kg algae, 73 kg invertebrates or 73 kg fish.

	Agricultural models					Marine model		
	Waste in the soil							
	0.07%		0.05%					
	via milk	via meat	via potatoes	via milk	via meat	via algae	via invertebrates	via fish
^{99}Tc	2	< 1	33	18	< 1	< 1	< 1	< 1
^{129}I ***	47	26	96	43	24	1	< 1	< 1
^{237}Np *	< 1	< 1	12	< 1	< 1	< 1	< 1	< 1
^{233}U	< 1	< 1	6	< 1	< 1	< 1	< 1	< 1
^{229}Th *	< 1	4	25	< 1	2	1	2	1
^{225}Ra *	14	2	41	14	2	< 1	< 1	< 1
^{226}Ra *	89	3	267	93	3	< 1	6	1
^{210}Pb	**	**	**	**	**	2	< 1	< 1

*   If radioactive equilibrium existed with all the daughter nuclides some of these daughter nuclides would have to be included in this table.

**   Equilibrium with ^{226}Ra is in this case improbable.

*** ^{129}I is supposed to be included in the waste. The influence of stable I present in the soil is not taken into account; if this were done and the specific activity of ^{129}I would be introduced, radiation doses due to ^{129}I would be lower.

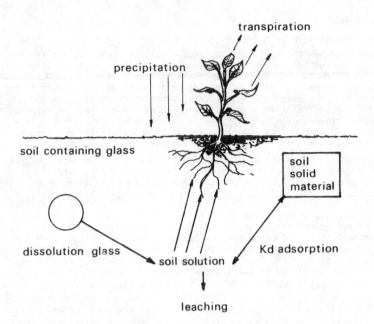

FIG. 3. Processes included in the description of the soil-plant transfer of radionuclides.

adsorbed by plant roots or can be leached by the surplus in precipitation (= precipitation - evapotranspiration). No selective uptake nor discrimination by the
roots is assumed, and also no redistribution of absorbed radionuclides inside the
plant. The concentration in the plant is calculated as the amount of water
transpired by the plant, multiplied with the concentration of the radionuclides
in the soil solution, divided by the amount of plant material produced.

External radiation model

Radionuclides in or on the soil contribute to the external radiation.
In the inhalation model 0.007% waste in the topsoil was assumed. In the agricultural models this was 0.07 and 0.005%. The radiation dose caused by the waste
at one metre above the soil (without shielding by a humus layer, by crops or by
irregularity of the surface) will then be 0.5, 5 and 0.4 rad per year. Somewhat
less than half of this amount is caused by fission products, the remainder from
the heavy metals and their daughters.

Marine model

If the waste reaches the surface under seawater, the radionuclides from
the waste can dissolve in this seawater, become diluted by currents, tides, etc.
Marine algae, invertebrates and fish can absorb these nuclides. People consuming
marine products (36.5 kg algae, 73 kg invertebrates or 73 kg fish per person per
year) can get a radiation dose from this which can be at the most 9% of the
ICRP-recommendations (see table II).

DISCUSSION

Sensitivity analysis

The following factors and parameters determine to a large extent the
results of the calculations.

As a rate of *diapyrism* 2.5 mm·a⁻¹ is chosen for a worst case analysis.
For lower rates the waste will reach the surface at a later time. For rates of
less than 0.5 mm·a⁻¹ groundwater will reach the waste, before this waste comes to
the surface. In that case transport of radionuclides to the biosphere will even be
more retarded by adsorption processes on solid material.

Landscape and biosphere and thus the dispersion of the waste is to a large extent determined by the *climate*. The influence of climate is difficult to quantify, but, in this study (a worst case analysis) minimal dispersion, and thus highest probable concentrations in soil, seawater and air are supposed.

The *transfer factors* grass - cattle - milk, grass - cattle - meat, seawater - marine products are taken from literature. The soil-plant transfer is calculated with a new approach using conservative factors. The degree of conservation will be shown later in this paragraph.

The amount of *slowly soluble compounds* in the rock salt is, in general, inversely proportional to the final radiation dose to man. In this study an average concentration of 10% slowly soluble compounds is assumed.

The *rate of dissolution of the glass* is, in general, proportional to the radiation dose to man. In this study a long term dissolution rate of 1.25×10^{-5} $g \cdot cm^{-2} \cdot a^{-1}$ is assumed, which means that glass spheres of 1 cm diameter are totally dissolved in 100 000 years and glass spheres of 1 mm diameter in 10 000 years.

By changing the *total amount of waste*, the *amount of radionuclides per volume glass*, or the *disposal concept* the radiation doses via the different pathways can be influenced.

Comparison to natural radioactivity

Although the calculated radiation doses to man in this worst case analysis are somewhat high, the situations which lead to the radiation doses, are comparable to situations in other parts of the world with a relatively high natural radioactive background (table III). This is the more so as the most critical radionuclide for ingestion by man is ^{226}Ra which is rather abundant in nature.

TABLE IIIa. Concentrations of ^{226}Ra in drinking water contaminated with waste, compared with a high concentration ^{226}Ra in Finnish drinking water.

	pCi ^{226}Ra$\cdot$litre^{-1}
Drinking water in Finland[a]	up to 256
Drinking water with 25 mg waste/m^3	3
Drinking water with 40 mg waste/m^3	5
Drinking water with 76 mg waste/m^3	9
Drinking water with 92 mg waste/m^3	11

TABLE IIIb. Concentration of ^{226}Ra in soils contaminated with waste, compared with an average soil and some geologic materials.

	pCi ^{226}Ra$\cdot$g^{-1}
Uranium ore, 40% U_3O_8[b]	100.000
Uranium ore, 0.25% U_3O_8[b]	700
Gold ore[c]	56
Rössing granite[c]	113
Average soil[d]	0.8
Soil with 0.07% waste	80
Soil with 0.005% waste	6

TABLE IIIc. Dose rate at 1 m above soils with waste compared with some high values measured at several places[d].

	rad$\cdot$a^{-1}
Kerala, India	1.1
Some places in Brasil	0.9 - 25
Ramsar, Iran	1.8 - 44
France	up to 1.8
Soil with 0.007% waste	0.5
Soil with 0.7% waste	5
Soil with 0.005% waste	0.4

[a] Kahlos, Asikainen (1973)
[b] Eisenbud (1975)
[c] Huwyler (1977)
[d] UNSCEAR (1977)

The conservatism of the calculations can be estimated by comparing the ^{226}Ra content of an average soil with the content of a soil with 0.005% waste. The latter soil contains 7.5 times as much ^{226}Ra as an average soil. The radiation dose, however, which is calculated for the soil with 0.005% waste is 225 times as high as average radiation dose caused by natural ^{226}Ra.

In table IV a summary is presented of the most important assumptions and of the results.

TABLE IV. Summary of the results and of the main assumptions.

Model	% Waste in soil/rock	Contaminated area	Estimate of the number of exposed people	Radiation dose as fraction of the ICRP-re-commendations*
Drinking water	0.07	0.5 km^2	10^3	< 0.5
Inhalation	0.007	5 km^2	10^1	< 0.1
External radiation I	0.007	5 km^2	10^1	< 1
Agricultural I	0.07	0.5 km^2		
milk			10^2	< 1
meat			10^1	< 1
External radiation II	0.07	0.5 km^2	10^1-10^2	<10
Agricultural II	0.005	100 km^2		
potatoes			10^5	< 3
milk			10^4	< 1
meat			10^3	< 1
External radiation III	0.005	100 km^2	10^3-10^5	< 1
Marine	-	25 km^2	no estimation	< 0.1

*for individual members of the public.

Remarks

a. All results are based on the assumption that the salt rises with a rate of 2.5 mm·a^{-1}, and the waste reaches biosphere after 250 000 years.

b. All people concerned stay all year round, day and night in the contaminated area or obtain all their food from this area.

c. Rocksalt contains 10% slowly soluble compounds.

d. At a depth of 600-680 m, 2500 m^3 waste is stored over an area of 600 x 800 m^2.

e. Waste in the form of glass dissolves in water at a rate of 1.25 x 10^{-5}g·cm^{-2}a^{-1}.

f. The glass is broken into spheres of 1 cm diameter in the drinking water model, agricultural mode I (+ external radiation II), into spheres of 1 mm diameter after erosion in agricultural model II (external radiation model I). In the marine model glass dissolves at once.

CONCLUSION

The highest radiation doses for individuals may arise if the contamination caused by the waste is confined to a small area. The radionuclides which contribute most to the internal radiation dose after 250 000 years, are the ones which are also present in the natural environment; the most important being ^{226}Ra. The concentration of ^{226}Ra in soil and water can reach the same order of magnitude as found in certain areas of the world due to natural radioactivity. Under the most unfavourable conditions, the radiation doses to people consuming contaminated food (agricultural or marine products) or water, or inhaling contaminated air, can reach a level which is still acceptable for individual members of the population, according to ICRP-standards. For a highly improbable combination of circumstances the radiation dose caused by external radiation could exceed, for a limited number of people, the ICRP-standards by at the most a factor of 10.

In a final site specific study based on *in situ* geohydrological measurements worst case radiation doses will be calculated, which are expected to be considerably lower than the doses calculated in this report.

ACKNOWLEDGEMENTS

This study has been carried out for the Stichting Energieonderzoek Centrum Nederland (ECN) under contract with the Commission of the European Communities (Contr. nr. 026-76-9 WAS N).

REFERENCES

Burkholder, H.C., M.A. Cloninger, D.A. Baker, G. Jansen: "Incentives for partitioning high level waste", BNWL-1927, U-70, Battelle Pacific North West Laboratories, Richland, U.S.A. (1975).

Van Dorp, F., R. Eleveld, M.J. Frissel, P. Poelstra: "Transport of radionuclides stored in a salt dome to and through the biosphere" (in Dutch), Association EURATOM-ITAL, Wageningen, The Netherlands (1978).

Van Dorp, F., R. Eleveld, M.J. Frissel: "A new approach for soil-plant transfer calculations", to be published, IAEA-SM-237/13 (1979).

Eisenbud, M.: "Environmental radioactivity", Academic Press New York, London (1975).

Huwyler, S.: "Probleme der Versorgung mit Natururan", Atomwirtschaft-Atomtechnik, 22, 577-580 (1977).

ICRP-2: "Recommendations of the International Commission on Radiological Protection", ICRP Publication 2, Report of Committee II on Permissible Dose for Internal Radiation, Pergamon Press, New York, (1959).

ICRP-6: "Recommendations of the International Commission on Radiological Protection" (As amended 1959 and Revised 1962) ICRP Publication 6, Pergamon Press, Oxford (1964).

ICRP-9: "Recommendations of the International Commission on Radiological Protection" (Adopted September 17, 1965) ICRP Publication 9, Pergamon Press, Oxford (1966).

Kahlos, H. and M. Asikainen: "Natural radioactivity of ground water in the Helsinki area", Institute of Radiation Physics, Helsinki, Finland, Report SFL-A19, (1973).

UNSCEAR: "Sources and effects of ionizing radiation", United Nations Scientific Committee on the Effects of Atomic Radiation, Report to the General Assembly, with annexes, United Nations, New York, No. E. 77 IX.1 (1977).

Discussion

P.P. de REGGE, Belgium

Looking at the tables in your paper, there is no reference to any plutonium isotope. Is it indeed your conclusion that, after 250.000 years, plutonium does not represent a significant hazard ?

F. van DORP, The Netherlands

After 250.000 years Pu has decayed so much that it represents only a very small risk compared to the total risk of the waste, less than 1 %.

S. ORLOWSKI, CEC

Dans votre conclusion, vous mentionnez un cas "hautement improbable", conduisant à des doses voisines de celles acceptées par l'ICRP. Qu'entendez-vous par "hautement improbable ?

F. van DORP, The Netherlands

The term "highly improbable" refers to a combination of many subsequent phenomena, of which each one is rather improbable. It is almost impossible to estimate the degree of improbability, but as the aim of our study was a worst case analysis, we considered a sequence and combination of improbable phenomena and processes like : a fast rise of the salt ; dissolution of the salt leaving the waste untouched and undisplaced ; no erosion of the waste ; and no sedimentation over the waste in a period of at least 30.000 years after the first waste appeared at the soil surface (after 250.000 years).

F. GIRARDI, CEC

I see that the discussion is going towards the more general subject of "risk analysis", which has been treated extensively in another workshop held jointly by NEA and CEC in Ispra in 1977. I would propose that the discussion is limited to problems on migration of long-lived radionuclides in the geosphere, which is the subject of this workshop, in order to be more productive in the few days that we have available.

DETERMINATION DES CONDITIONS DE TRANSFERT
DE PRODUITS RADIOACTIFS DANS UN MASSIF GRANITIQUE FISSURE
AU MOYEN D'ESSAIS IN SITU ET D'ESSAIS
SUR ECHANTILLONS

P. Calmels[*], J. Guizerix[*]
P. Peaudecerf[**] et J. Rochon[**]

[*] Commissariat à l'Energie Atomique - Centre d'Etudes Nucléaires de Grenoble
[**] Bureau de Recherches Géologiques et Minières

Résumé

        Dans le cadre du programme gestion et stockage des déchets radioactifs
de la Commission des Communautés Européennes, ont été étudiées les modalités de
transfert des produits au sein des massifs granitiques fissurés. Dans un premier
temps, une étude expérimentale sur un massif granitique et sur échantillons a
permis de rassembler des informations importantes sur les circulations dans les
fissures et les adsorptions à attendre pour certains produits de fission, comme
le strontium et le cesium.

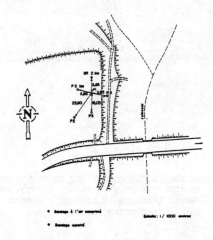

Figure 1 - Plan de situation des forages

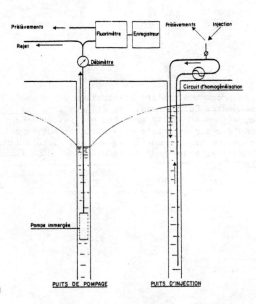

SCHÉMA DE PRINCIPE DE LA MESURE   PL.1

Figure 2 - Schéma de principe de la mesure des caractéristiques
de transfert

1. INTRODUCTION

Les travaux de recherche que nous présentons ici ont été réalisés sous contrat à frais partagés entre la Commission des Communautés Européennes et le Commissariat à l'Energie Atomique dans le cadre du programme d'action indirecte Gestion et stockage des déchets radioactifs.

Le stockage dans un massif cristallin exige que l'on connaisse les modalités de transfert des substances au sein de ce type de massif. Ces éventuelles migrations se feront en suivant les fissures naturelles de la roche ou bien celles qui seront provoquées par l'élévation de la température locale due à l'énergie rémanente des produits stockés ou encore par la décompression de la roche au voisinage de l'excavation.

Mais, si l'on connaissait assez bien les caractéristiques des migrations dans les milieux poreux granulaires, celles relatives aux milieux fissurés cristallins étaient hypothétiques et il convenait de les reconnaître sur le milieu naturel en place. La question précise qui se posait était de savoir si les techniques de mesure utilisées dans les milieux granulaires, les bases théoriques des interprétations de ces mesures et celles des modèles de prévision étaient applicables aux milieux cristallins fissurés, ou, de façon plus réaliste, quelles sont les modifications que l'on doit faire subir aux unes et aux autres pour les adapter aux particularités des milieux cristallins.

Cette recherche est menée en collaboration avec l'Ecole des Mines de Paris et le Commissariat à l'Energie Atomique. Dans l'exposé antérieur (ref. 2) "Représentation sur modèle de la migration des radioéléments dans les roches fissurées", MM. GOBLET, LEDOUX, de MARSILY et BARBREAU traitent le problème de la modélisation des transferts. Dans cette communication, nous présentons plus particulièrement la phase expérimentale de l'étude en mettant en évidence les résultats auxquels on a été conduit.

2. DESCRIPTION SOMMAIRE DU DISPOSITIF EXPERIMENTAL

Dans le massif granitique choisi, ont été exécutés six forages de 175 mm de diamètre jusqu'à 18 m, puis 115 mm de 18 à 40 m environ (P2 excepté) suivant le schéma de la figure 1. Les forages satellites sont à 12 m du forage central P1. La reconnaissance préliminaire et l'examen des carottes ont montré que les dix premiers mètres avaient subi une fissuration horizontale et une altération météorique. En profondeur, le granite est sain et l'on observe une fissuration principalement verticale.

- Le puits central P1 est muni d'une pompe immergée et d'un débit-mètre à flotteur magnétique permettant la mesure et l'enregistrement des débits.

- Sur tous les forages, un dispositif de mesure des niveaux d'eau permet de suivre les variations de niveau dans une gamme de temps très étendue, de la fraction de seconde à la dizaine de jours. Il est basé sur l'utilisation de transmetteurs pneumatiques de niveau (bulle à bulle) et d'enregistreur rapide.

- Dans chaque piézomètre, un circuit de pompage provoque la circulation des eaux sur toute la hauteur du forage afin d'homogénéiser les concentrations durant toute la durée de l'injection des traceurs (cf. figure 2).

- Des prélèvements automatiques ou manuels sur la conduite issue du puits central permettent de suivre la composition des eaux pompées.

3. PRINCIPE DES ESSAIS

L'objectif des essais est d'étudier les écoulements dans le massif granitique expérimental, d'en proposer un schéma explicatif le plus simple possible et d'évaluer les valeurs à attribuer aux paramètres physiques retenus.

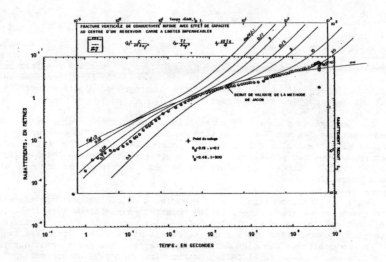

Figure 3 - Pompage d'essai du 16 au 29/11/1977. Rabattement au
puits de pompage (débit : Q=2,2 m³/h)

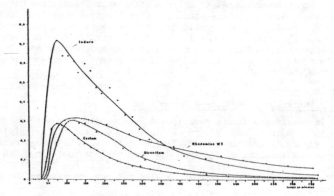

Figure 4 - Comparaison des courbes de restitution de divers produits
sur P4-P1 (12m) (non déconvoluées de la fonction d'injection)

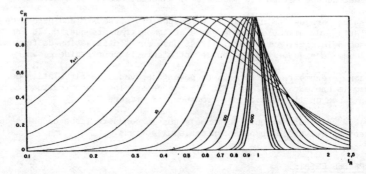

Figure 5 - Abaque pour une injection instantanée dans un écoulement
convergent

Un pompage à débit constant (2 m³/h environ) est maintenu dans le puits central P1 durant une quinzaine de jours. Ainsi le schéma hydraulique est grossièrement radial convergent.

- D'une part, on relève les variations des charges d'eau au sein du massif dans chacun des forages. L'interprétation de ces variations en régime transitoire conduit à évaluer les paramètres hydrodynamiques.

- D'autre part, durant le pompage, des injections brèves de traceurs sont effectuées dans les piézomètres satellites. Leur restitution dans l'eau prélevée permet d'apprécier les paramètres de transfert dans chacune des directions cardinales. Ont été injectées des substances réputées bon traceur du mouvement de l'eau et certains produits témoins, comme des sels de strontium et de césium.

- Enfin, des essais de migration sur échantillons de granite fissuré prélevés à 30 cm de profondeur ont permis d'évaluer l'importance des fixations des produits injectés in situ.

D'une façon générale, ces essais ont confirmé une forte hétérogénéité apparente du milieu : des liaisons privilégiées sont apparues (P4 - P1), tant du point de vue hydrodynamique que cinématique.

4. INTERPRÉTATION DES DONNÉES EXPÉRIMENTALES

4.1. Interprétation des variations de niveaux

Celles-ci ont été conduites par application de différents schémas théoriques de l'hydraulique souterraine près des puits.

La partie initiale des courbes de rabattements en fonction du temps est caractéristique d'effets de capacité du puits (cf. ref. 1) ; les dernières parties traduisent l'influence de limites imperméables pour le réservoir. Quant à la partie médiane, elle peut être assimilée aux effets d'une fracture verticale selon la schématisation de A.C. GRINGARTEN (ref. 3 et 4), comme le montre la figure 3. Mais on peut aussi expliquer d'une façon approchée ces comportements comme ceux d'un milieu continu homogène et anisotrope (cf. ref. 2). La perméabilité moyenne équivalente est de l'ordre de 10^{-6} m/s ou 0,1 Darcy.

Quelle que soit l'hypothèse choisie, on doit supposer l'existence de limites imperméables situées à 250 m environ du dispositif d'essai.

4.2. Interprétation des transferts de masse

4.2.1. Essais in situ

De nombreux traçages ont été effectués à partir des piézomètres satellites afin d'évaluer les caractéristiques de transfert. Ils ont confirmé l'hétérogénéité apparente du massif expérimental.

Les données issues de ces essais sont de trois types :

- les courbes de disparition des substances dans les divers forages d'injection.

- les taux de restitution au puits de pompage P1,

- la forme des courbes de restitution lorsque le traceur réapparaît (cf. fig .4).

Les réponses impulsionnelles relevées au puits P1 ou dans certains cas déduites de la courbe de restitution par déconvolution de la fonction d'entrée au puits d'injection sont interprétées par superposition sur les abaques établis par J.P. SAUTY (cf. ref. 6), comme le montre le rapprochement des figures 5 et 6.

Pour cela, on fait l'hypothèse que les écoulements sont radiaux convergents et le milieu isotrope, ce qui constitue ici une réelle distorsion de la réalité. Alors, l'équation différentielle donnant la concentration moyenne à la distance r du puits en pompage est :

$$\frac{\partial \overline{c}}{\partial t} = D \frac{\partial^2 \overline{c}}{\partial r^2} - u \frac{\partial \overline{c}}{\partial r} \qquad (1)$$

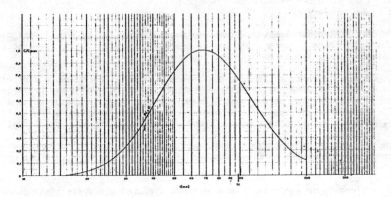

Figure 6 - Courbe de restitution du chlorure (Cl⁻)
(après déconvolution) trajet P4-P1

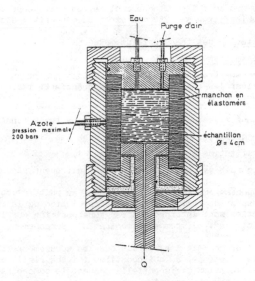

Figure 7 - Schéma de la cellule de percolation

avec u, vitesse d'entrainement à la distance r,
soit :

$$u = \frac{Q}{2\pi rh\omega} \text{ soit, si on pose } A = \frac{A}{2\pi h\omega}, \ u = \frac{A}{r}$$

h, épaisseur de l'aquifère
ω, porosité cinématique.

Le coefficient de dispersion longitudinale D est aussi égal pour ces vitesses d'écoulement au produit D = αu, α étant la dispersivité du milieu (diffusion moléculaire négligeable), et l'on peut écrire (1) sous la forme :

$$\frac{\partial \overline{C}}{\partial t} = \frac{A}{r} \frac{\partial^2 \overline{C}}{\partial r^2} - \frac{A}{r} \frac{\partial \overline{C}}{\partial r} \tag{2}$$

Si, de plus, se superposent des effets d'échanges physico-chimiques avec la roche, l'équation (2) devient :

$$\frac{\partial}{\partial t}\left(\overline{C} + \frac{1-\omega}{\omega} \rho_s F\right) = \alpha \frac{A}{r} \frac{\partial^2 \overline{C}}{\partial r^2} - \frac{A}{r} \frac{\partial C}{\partial r} \tag{3}$$

F est la concentration dans la phase immobile
et ρ_s sa masse volumique.

Dans le cas idéal d'une adsorption linéaire, réversible et instantanée,

$$F = K_d \overline{C}$$

et (3) peut s'écrire :

$$\frac{\partial \overline{C}}{\partial t} = \frac{\alpha}{r} \frac{A}{R} \frac{\partial^2 \overline{C}}{\partial r^2} - \frac{1}{r} \frac{A}{R} \frac{\partial C}{\partial r} \tag{4}$$

R est le coefficient de retard, et l'équation (4) est identique à (2), en remplaçant A par A/R ou la porosité cinématique ω par une porosité cinématique apparente $\omega_- = R\omega$.

Ainsi, dans ces hypothèses, si l'on superpose les courbes expérimentales C(t) sur les courbes théoriques du schéma hydrodispersif pur (fig. 5), la valeur apparente de la porosité peut nous éclairer sur l'importance des fixations.

Les résultats pour les traceurs du mouvement de l'eau sont :

TABLEAU I

Points d'injection	Traceurs	Vitesse moyenne (m/h)	Porosité cinématique ω	Dispersivité α (m)	Taux de restitution (%)
P4	Cl$^-$	7,3	$1,9 \cdot 10^{-4}$	1,5	75
P4	I$^-$	7,6	$1,9 \cdot 10^{-4}$	0,8	72
P4	Rhodamine Wt	6,5	$2,3 \cdot 10^{-4}$	0,6	38
P2bis	NO$_3^-$	0,12	$1,2 \cdot 10^{-2}$	4,0	73
SR	I$^-$	0,064	$2,2 \cdot 10^{-2}$	0,6	31
P3	Rhodamine Wt	(non apparue - vitesse trop lente)			

et pour les substances témoin, strontium et césium, :

P4	Sr^{++}	6,5	$2,3 \cdot 10^{-4}$	0,6	31
P4	Cs$^+$	7,9	$1,9 \cdot 10^{-4}$	0,6	29

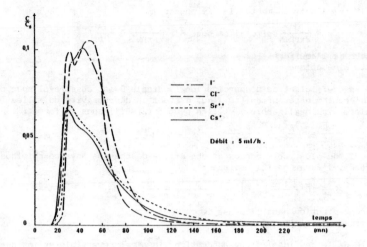

Figure 8 - Essais sur échantillons de roche
Courbes de restitution

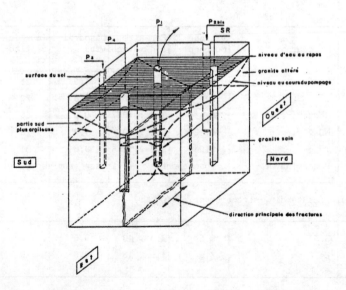

Figure 9 - Schématisation hydrodynamique du massif expérimental

4.2.2. Essais sur échantillons de roche

Afin de mieux comprendre les réactions entre les substances et la roche, notamment les matériaux de remplissage des fissures, des essais de transfert ont été réalisés sur échantillons prélevés sur la carotte du forage de reconnaissance SR2 entre 29 et 30 m de profondeur. Ces échantillons cylindriques de petites dimensions (longueur et diamètre de l'ordre de 4 cm) sont chacun parcourus par une fissure dont le plan principal est parallèle à leur axe.

Des analyses par diffraction aux rayons X des matériaux d'altération tapissant la fissure ont mis en évidence les substances suivantes :
- plagioclase
- montmorillonite
- quartz
- muscovite
- hydroxyde de fer (goethite).

L'eau utilisée pour les déplacements était celle prélevée in situ. Les concentrations et les vitesses sont comparables à celles des essais in situ. Les échantillons sont placés dans une cellule dont le schéma est indiqué à la figure 7. Les courbes de restitution sont données figure 8.

Les courbes de restitution de l'échantillon 2 ont conduit à une interprétation en schéma bicouche. Les résultats sont rassemblés ci-après :

TABLEAU II - Echantillon n° 1

Substance	Vitesse u (cm/h)	Porosité cinématique	Dispersivité α (m)	Taux de restitution(%)
NO_3^-	13,4	$2,9 \cdot 10^{-2}$	$8 \cdot 10^{-4}$	près de 100
Na^+	7,2	$5,5 \cdot 10^{-2}$	$3 \cdot 10^{-3}$	64
RhWt	11,5	$3,5 \cdot 10^{-2}$	$2,5 \cdot 10^{-3}$	47
Uranine	4,5	$8,7 \cdot 10^{-2}$	$5 \cdot 10^{-3}$	57

TABLEAU III - Echantillon n° 2

Substance	Vitesse u (cm/h)	Porosité cinématique	Dispersivité α (m)	Taux de restitution(%)
I^-	8,0 et 4,8	$4,9 et 8,2 \cdot 10^{-2}$	4 et $11 \cdot 10^{-4}$	77
Cl^-	8,6 et 5,0	$4,6 et 7,9 \cdot 10^{-2}$	11 et $7 \cdot 10^{-4}$	79
Sr^{++}	8,7 et 5,0	$4,6 et 7,9 \cdot 10^{-2}$	7 et $40 \cdot 10^{-4}$	62
Cs^+	9,3 et 5,1	$4,3 et 7,8 \cdot 10^{-2}$	11 et $22 \cdot 10^{-4}$	60

4.2.3. Principaux résultats des essais de transfert

L'étude expérimentale a permis de rassembler des données sur les migrations en milieu fissuré. Ces données pourraient être complétées en étudiant le transfert d'autres produits. Nous nous sommes limités ici aux traceurs du mouvement de l'eau, au césium et au strontium. On pourrait envisager d'étudier d'autres produits sur les mêmes milieux.

Les résultats des mesures pourraient aussi être analysés avec plus de précision afin d'expliquer le mieux possible la forme des courbes de restitution. Une interprétation plus poussée est présentée à la référence 2. Mais on doit noter à ce sujet que les concentrations utilisées dans nos expériences sont beaucoup plus importantes que celles que l'on trouvera dans l'environnement d'un stockage de produits radioactifs et que les phénomènes d'échanges pourront être fortement différents.

La première interprétation présentée ici permet d'avancer les résultats suivants :

1. Les anions testés, I^-, Cl^- et NO_3^-, ne montrent pas d'adsorption mesurable. Les cations et les composés organiques subissent eux les effets d'interactions avec le milieu naturel.

2. Les paramètres hydrocinématiques du milieu, et principalement la porosité ciné-matique, mesurables par le transfert des anions, apparaissent répartis de fa-çon anisotrppe autour du puits P1 suivant l'orientation des fissures.

3. Le strontium, à ces concentrations, a un taux de restitution de 30% environ pour la durée de l'essai ; il subit de la part du milieu une adsorption de forme particulière : la désorption est très lente.

4. Le césium par contre subit apparemment une fixation partielle et pratiquemment irréversible pendant la durée des expériences.

CONCLUSION

La présente étude expérimentale a montré la complexité de l'hydrodynami que locale d'un massif granitique fissuré. Le passage à la phase de prévision exige une schématisation du système. Celle-ci doit être choisie au mieux, mais différentes possibilités demeurent. Nous donnons à la figure 9 une schématisation possible où l'on distingue deux niveaux :

- une partie superficielle caractérisée par une formation importante et relative-ment isotrope,

- une partie profonde où le granite est massif et dans lequel se développe une fissuration verticale est-ouest, bien visible sur la carotte prélevée en SR.

On remarquera dans la communication évoquée plus haut (ref. 2) une in-terprétation un peu différente.

REFERENCES

/1/ PAPADOPULOS (I.S.), COOPER (H.H.) Jr. 1967 .- Drawdown in a well of large dia-meter .- *Water Resources Research, 3 : p. 241-244.*

/2/ GOBLET (P.), LEDOUX (E.) de MARSILY (G.), BARBREAU (A.) .- Représentation sur modèle de la migration des radioéléments dans les roches fissurées .- *Workshop on the migration of long-lived radionuclides in the geosphère, OECD-CCE, Brussels, 1979.*

/3/ GRINGARTEN (A.C.), RAMEY (H.J.) Jr., RAGHAVAN (R.) .- Unsteady state pressure distributions created by a well with a single infinite conductivity vertical fracture .- *Soc. Pet. Eng. J. (aug. 1974) p. 347-360.*

/4/ EARLOUGHER (R.C.) Jr. .- Advances in well test analysis .- *New York ; Dallas : Soc. of Pet. Eng. of AIME, 1977, 264 p.*

/5/ BEAR (J.) .- Dynamics of fluids in porous media .- *New York, American Alsevier, 1972.*

/6/ SAUTY (J.P.) .- Contribution à l'identification des paramètres de dispersion dans les aquifères par interprétation des expériences de traçage .- *Thèse doct.-ing., Grenoble, 1977.*

/7/ PEAUDECERF (P.), SAUTY (J.P.) .- Application of a mathematical model to the characterization of dispersion effects on groundwater quality .- *IAWPR Sotckholm, conférence 1978.*

REPRESENTATION SUR MODELE DE LA MIGRATION DES RADIOELEMENTS

DANS LES ROCHES FISSUREES

P. Goblet, E. Ledoux, G. de Marsily
Ecole des Mines de Paris,
Centre d'Informatique Géologique
Fontainebleau (France)

A. Barbreau
Commissariat à l'Energie Atomique
Institut de Protection et Sûreté Nucléaire
Fontenay aux Roses (France)

La migration des éléments en solution dans les roches fissurées est abordée en décrivant le milieu fissuré par le biais du milieu continu équivalent, hétérogène et anisotrope. La rétention des substances est représentée par des mécanismes d'adsorption limités aux plans des fissures.

Après un exposé du principe de cette modélisation, il est donné un exemple d'interprétation d'expériences de traçage réalisées sur un massif granitique fissuré par le Bureau de Recherches Géologiques et Minières et le Centre d'Etudes Nucléaires de Grenoble. Ce travail a été réalisé pour le compte de la Commission des Communautés Economiques Européennes et du Commissariat à l'Energie Atomique dans le cadre d'un contrat à frais partagés concernant le programme d'action indirecte Gestion et Stockage des Déchets Radioactifs.

La présente communication fait état des résultats acquis à l'occasion d'un travail entrepris pour le compte de la Commission des Communautés Economiques Européennes et du Commissariat à l'Energie Atomique, dans le cadre d'un contrat à frais partagés concernant le programme d'action indirecte Gestion et Stockage des Déchets Radioactifs.

Ce travail, réalisé en collaboration avec le Bureau de Recherches Géologiques et Minières et le Centre d'Etudes Nucléaires de Grenoble, a porté sur la représentation sur modèle mathématique de la migration des radioéléments dans les roches fissurées.

Après un aperçu théorique sur le principe et les méthodes employés pour la modélisation, la représentativité des modèles sera testée à la faveur d'essais sur échantillons et in situ réalisés sur un massif de granite fissuré, avec différentes substances en solution.

PRINCIPE DE LA MODELISATION

Le mécanisme de migration des radioéléments en solution dans un milieu fissuré peu perméable résulte de la combinaison de plusieurs phénomènes physiques liés à l'écoulement de l'eau dans les fissures d'une part, et aux interactions entre les substances dissoutes et la matrice rocheuse d'autre part.

La mise en équations des mécanismes considérés comme les plus significatifs, puis la résolution de ces équations par une méthode numérique constituent le modèle mathématique dont le principe va être décrit.

Hypothèses du modèle

Compte-tenu de l'échelle à laquelle se posent à long terme les problèmes de migration (entre le cimetière et la surface), les hypothèses suivantes ont été adoptées:

a) Hypothèses concernant l'écoulement de l'eau:

- hypothèse du milieu continu équivalent; les écoulements seront étudiés à l'échelle du Volume Elémentaire Représentatif qui intègre l'hétérogénéité fondamentale des milieux naturels (pores ou réseau de fissures);
- hypothèse du milieu hétérogène et anisotrope, les directions principales d'anisotropie pouvant varier dans l'ensemble du milieu;
- hypothèse du réseau d'écoulement non uniforme, pouvant présenter des variations importantes de vitesse d'écoulement au voisinage de points particuliers (sources, sites de stockage);
- hypothèse du régime d'écoulement permanent.

b) Hypothèses concernant la migration des radioéléments:

- déplacement par l'entrainement naturel de l'écoulement, appelé convection;
- déplacement diffusif sous l'effet de la diffusion moléculaire, de la dispersion cinématique et de la dilution dans la porosité cinématique du milieu en supposant la masse volumique de la solution invariante avec la concentration (hypothèse du traceur);
- interaction entre un soluté et la matrice minérale suivant différents types de loi de rétention indépendamment des divers corps en présence dans la solution (adsorption-désorption non sélective).

Equations

Sous les hypothèses précédentes, la migration des radioéléments est représentée par les équations suivantes:

a) Equation de la dispersion généralisée:

$$\text{div } D \text{ (grad C)} - \text{div (VC)} = \omega \left(\frac{\partial C}{\partial t} + \lambda C\right) + \mu \left(\frac{\partial F}{\partial t} + \lambda F\right)$$

b) Loi de rétention:

$$\frac{\partial F}{\partial t} = K_1 \left(C - \frac{F}{K_d}\right) \qquad \text{avec} \qquad \begin{array}{l} K_1 = K_1 \text{ ads lors de l'adsorption} \\ K_1 = K_1 \text{ dés lors de la désorption} \end{array}$$

avec les notations suivantes:

C : concentration en un radioélément dans la solution $[M][L]^{-3}$
F : concentration retenue sur la paroi des fissures $[M][L]^{-2}$
V : vitesse de Darcy de l'écoulement
$D = \alpha|V|$: coefficient de dispersion (tensoriel) $[L]^2[T]^{-1}$
α : dispersivité intrinsèque (tensorielle) $[L]$
ω : porosité cinématique (sans dimension)
μ : densité de fracturation $[L]^{-1}$
λ : coefficient de décroissance radioactive $[T]^{-1}$
K_d : coefficient de distribution de fracture $[L]$

K_1 ads: coefficient cinétique d'adsorption $[L][T]^{-1}$
K_1 dés: coefficient cinétique de désorption $[L][T]^{-1}$

La loi de rétention choisie est de type linéaire, avec cinétique linéaire différente lors de l'adsorption et de la désorption; elle suppose la connaissance de trois paramètres, un paramètre statique K_d et deux paramètres cinétiques K_1 ads et K_1 dés.

Un cas particulier intéressant est obtenu lorsque K_1 ads et K_1 dés tendent vers l'infini. La loi de rétention devient alors linéaire et instantanée et prend la forme:

$$F = K_d C$$

Nous admettrons cette formulation pour la suite dans le cas des substances qualifiées de "bons traceurs", c'est à dire faiblement et instantanément échangeables.

Méthodes numériques

Considérant la complexité des équations à résoudre, on a eu recours à une méthode numérique de résolution: méthode en différences finies dans le cas à une dimension, méthodes en éléments finis isoparamétriques linéaires dans les cas à deux ou trois dimensions.

INTERPRETATION DES ESSAIS REALISES SUR UN MASSIF GRANITIQUE

Ces essais réalisés sur un massif granitique fissuré à faible profondeur par le BRGM et le CENG comportent principalement:

- des essais en laboratoire sur deux échantillons comportant une fissure ouverte; ils ont permis la détermination du type de loi de rétention entre la matrice minérale et différentes substances;

- des essais de traçage in situ réalisés en pompant de l'eau dans un forage central et en injectant des traceurs sur des piézomètres périphériques distants d'un douzaine de mètres.

FIG. 1 - ESSAIS SUR ECHANTILLONS

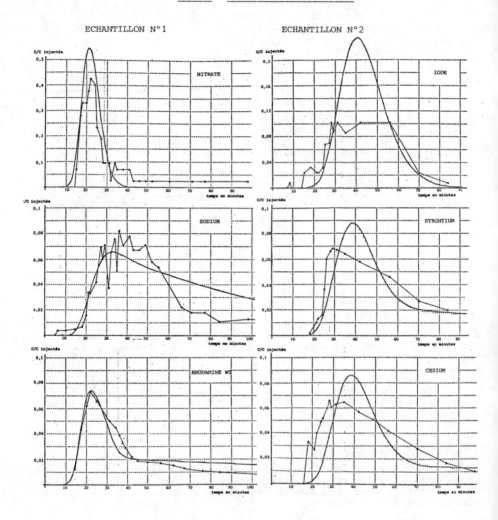

ECHANTILLON N°1   ECHANTILLON N°2

Les substances étudiées sont: NaI, NaNO$_3$, NaCl, Uranine, Rhodamine WT, SrCl$_2$, Cs$_2$SO$_4$.

Interprétation des essais sur échantillons

L'interprétation porte sur la comparaison entre la courbe de concentration mesurée à la sortie de l'échantillon ayant subi une injection brève d'une substance et la courbe calculée au moyen du modèle monodimensionnel.

La démarche suivante a été adoptée. On choisit en premier lieu la substance présentant la vitesse de migration et le taux de restitution les plus élevés (ici NO$_3^-$, I$^-$ ou Cl$^-$). L'ajustement du modèle fournit alors la dispersivité et la porosité cinématique de l'échantillon vis à vis de la substance retenue comme "bon traceur". On interprète ensuite la migration plus lente des autres corps en supposant qu'ils donnent lieu à des phénomènes d'adsorption sur les parois de la fissure tout en conservant la même dispersivité et la même porosité cinématique considérées comme caractéristiques de l'échantillon.

L'identification des paramètres a été réalisée automatiquement à l'aide d'une méthode d'optimisation non linéaire.

Quelques-uns parmi les résultats obtenus sont consignés dans le tableau I et la figure 1.

TABLEAU I

Valeurs des paramètres de transfert obtenus sur échantillons

	Substance	Dispersivité (m)	Porosité cinématique	K$_d$ (m)	K$_1$ ads (m/h)	K$_1$ dés (m/h)
Echantillon n°1	NO$_3^-$ "bon traceur"	1,2.10^{-3}	2,6.10^{-2}	–	–	–
	Na$^+$			5,8.10^{-4}	9,4.10^{-3}	6,1.10^{-4}
	Rhodamine WT			8,6.10^{-4}	3,8.10^{-3}	2,5.10^{-4}
Echantillon n°2	Cl$^-$ I$^-$ "bons traceurs"	1,6.10^{-3}	6,5.10^{-2}	–	–	–
	Strontium			5,1.10^{-3}	1,7.10^{-3}	2,8.10^{-2}
	Cesium			7,0.10^{-3}	1,7.10^{-3}	3,4.10^{-2}

Interprétation des essais in situ

Le problème est ici double, car l'interprétation doit porter simultanément sur des données hydrodynamiques et des données de traçage. Compte-tenu des informations disponibles sur le site (niveaux dans cinq piézomètres), on a considéré le milieu fissuré comme homogène et anisotrope, admettant la direction moyenne du transfert, entre le puits de pompage et le piézomètre d'injection, comme direction principale d'anisotropie. L'interprétation a alors été conduite au moyen du modèle de transfert à deux dimensions.

Dans ces conditions, les paramètres moyens suivants, caractérisant l'aquifère ont pu être identifiés par une méthode analytique.

T_1 = 2,25.10^{-3} m^2/s transmissivité suivant la direction moyenne du transfert

T_2 = 4,5.10^{-6} m^2/s transmissivité suivant la direction orthogonale

S = 4,5.10^{-3} coefficient d'emmagasinement.

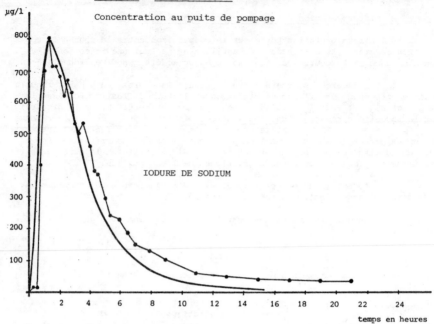

FIGURE 2 - ESSAIS.IN SITU

Concentration au puits de pompage

IODURE DE SODIUM

temps en heures

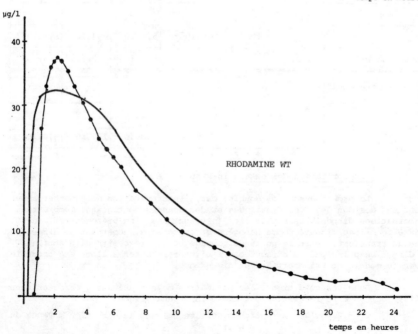

RHODAMINE WT

temps en heures

Rapport d'anisotropie $T_1/T_2 = 500$

L'ajustement du modèle de transfert sur les courbes de restitution a été conduit en recherchant, comme pour les essais de laboratoire, la dispersivité et la porosité cinématique correspondant à la substance présentant la plus grande vitesse de transfert (ici NaI), considérée comme "bon traceur". On a ensuite pour les autres substances utilisé des lois de rétention du même type que pour les essais de laboratoire dont on a tenté de déterminer les paramètres.

Les résultats obtenus font l'objet du tableau II et de la figure 2.

TABLEAU II

Valeurs des paramètres de transfert obtenus sur les essais in situ

Substance	Dispersivité (m)	Porosité cinématique	K_d (m)	K_1 ads (m/h)	K_1 dés (m/h)
I⁻ "bon traceur"			-	-	-
Rhodamine WT	3	$7,6.10^{-4}$	$8,6.10^{-3}$	3.10^{-3}	$2,5.10^{-4}$

Remarquons que l'Iodure de Sodium, considéré comme bon traceur, n'est restitué qu'à 72 % au cours de l'expérience. Cette observation ne pouvant être prise en compte par une augmentation de la dispersivité, il a fallu admettre l'existence d'une rétention pour cette substance que l'on a supposée instantanée et irréversible, tout comme si seulement 72 % de la masse initiale avait été effectivement injectée.

De plus, dans le cas des expériences in situ, il existe un paramètre supplémentaire représenté par la densité de fracturation μ ; cette propriété est en effet fonction de la fréquence des fissures et non pas de leur ouverture. Il n'est donc pas possible de la relier à la porosité cinématique.

L'ajustement du modèle conduit à adopter une valeur de 2 m^{-1} pour la densité de fracture, ce qui semble compatible avec les observations faites sur le milieu.

On constate que le passage des échantillons au terrain a nécessité une modification notable du coefficient de distribution K_d, mais pas des vitesses d'adsorption-désorption. Cette observation pourrait être justifiée par les modifications des minéraux tapissant les fissures des carottes à la suite des lavages successifs auxquelles elles ont été soumises.

CONCLUSION

Les résultats qui viennent d'être présentés suggèrent les conclusions suivantes:

- Une loi de rétention à trois paramètres (un paramètre statique et deux paramètres cinétiques) permet de rendre compte de l'allure des phénomènes de transfert aussi bien sur les échantillons que sur le terrain.

- La méthodologie d'ajustement du modèle, basée sur le calage préalable de la porosité cinématique et de la dispersivité associées à un couple terrain-"bon traceur" paraît intéressante, dans la mesure où elle permet de diminuer le nombre de paramètres à déterminer simultanément.

- Les paramètres cinétiques des lois de rétention obtenus sur échantillons continuent à être valables in situ. Cette constatation nous a permis de limiter finalement le nombre de paramètres à deux, à savoir la densité de fracturation et le coefficient de distribution.

- Malgré la méthodologie proposée, le nombre de paramètres restant important, il est préférable d'avoir recours à une méthode automatique d'identification. Ceci n'a été, pour le moment, possible que dans le cas des essais sur échantillons, pour lesquels les durées de calcul des simulations étaient de l'ordre de quelques minutes.

- Les valeurs des paramètres du transfert obtenus sont conditionnées par la détermination des paramètres hydrodynamiques du milieu, donc en fin de compte, par les hypothèses faites sur le réseau d'écoulement. Il importe donc de soigner en tout premier lieu les observations et les interprétations hydrogéologiques.

Ces conclusions ne s'appliquent qu'à un seul site expérimental; il serait approprié de les vérifier à la faveur d'autres expériences en faisant varier notamment les vitesses de transfert, ainsi que les distances parcourues. La possibilité de déterminer les paramètres cinétiques des lois de rétention sur échantillon demandent de plus à être confirmée sur d'autres carottes.

Discussion

F. van DORP, The Netherlands

If I understand rightly, you attribute the tails of your percolation curves to different adsorption and desorption rates. These tails, however, could also be explained by dead end pores with more or less immobile water.

E. LEDOUX, France

Ce phénomène est pris en compte par la loi de rétention. Cette loi exprime en effet une relation dépendante du temps entre la concentration de la solution constituée par l'eau qui circule (contenue dans la porosité cinématique) et la concentration retenue dans les fissures et dans l'eau immobile. La loi de rétention réalise probablement une approche globale de nombreux mécanismes complexes.

L.H. BAETSLE, Belgique

Pourquoi n'a-t-on pas utilisé le tritium comme traceur ?

E. LEDOUX, France

Il fallait se libérer au maximum des difficultés d'expérimentation, ce qui a conduit à éliminer les substances radioactives pour le traçage.

L.H. BAETSLE, Belgique

Quelle est l'extrapolabilité du coefficient μ à d'autres sites ?

E. LEDOUX, France

Le coefficient μ, densité de fracturation, égal à 2 m^{-1} dans l'exemple de notre site expérimental, n'est certainement pas applicable à d'autres sites ; il est représentatif des conditions moyennes de fissuration à l'échelle de l'expérience (15 m).

D. RAI, United States

Do I understand right that in the initial model the effect of the solution composition (radiotracer and other competing and complexing ions) on adsorption-desorption is not considered ? Solution composition has a tremendous effect on adsorption-desorption. Will the effect of solution composition be considered in future experiments ?

E. LEDOUX, France

En effet, l'influence de la composition de la solution sur la loi de rétention n'a pas été étudiée. Afin de limiter au maximum cet effet on a opéré à de très faibles concentrations. Monsieur Rochon (BRGM-SGN) pourra sans doute fournir un complément d'information à ce sujet.

J. ROCHON, France

Les interactions entre le sous-sol et les radioéléments sont trop complexes pour être représentées par le simple coefficient de distribution Kd. Outre l'adsorption, peuvent exister les phénomènes de précipitation, de complexation, les phénomènes de partage (équilibre de DONNAN), l'échange d'ions etc. La loi régissant l'équilibre n'est pas linéaire, même aux faibles concentrations.

Pour des essais d'interprétation d'expériences en colonne capillaire, nous avons utilisé pour simuler l'interaction :

- une isotherme de type LANGMUIR pour l'adsorption,

- la loi d'action de masse pour l'échange d'ions.

Les premiers résultats que nous présentons montrent que les ajustements sont très satisfaisants (couple Cs-quartz).

D'autres essais effectués avec du Tc sur de la sidérite peuvent être simulés en ajoutant un terme de consommation à l'équation de dispersion.

Nous insistons sur le fait que la rétention de chaque élément sur un minéral est spécifique et qu'il n'est pas possible de trouver une loi générale valable pour tous les éléments dans toutes les conditions.

G. LENZI, Italy

Avez-vous considéré le paramètre de la granulométrie sur le quartz en fonction de l'adsorption de radioéléments dans la colonne, dans les essais que vous avez faits en laboratoire ?

J. ROCHON, France

Tous nos minéraux ont été broyés à une granulométrie inférieure à 80 µm.

P. GLASBERGEN, The Netherlands

I noticed that there was a large difference in the height and the length of the elements in your grid. Are there no problems with the solution of your numerical equations ?

E. LEDOUX, France

L'allongement des éléments du maillage "éléments finis" dans le sens du transfert des substances a justement pour but de réduire les problèmes de résolution numérique des équations. Compte tenu du rapport d'anisotropie déterminé pour les perméabilités ($Kx/Ky \simeq 500$) il a semblé nécessaire de dilater les éléments afin de ramener les flux traversant le grand côté et le petit côté des rectangles à des valeurs comparables.

Session II
Radionuclides Migration and Retention

Chairman - Président
M. A.F. BARBREAU
(France)

Séance II
Migration et rétention des radionucléides

PRELIMINARY RESULTS ON COMPARISON OF ADSORPTION-DESORPTION METHODS AND STATISTICAL TECHNIQUES TO GENERATE Kd PREDICTOR EQUATIONS

R. J. Serne, Dhanpat Rai, and J. F. Relyea
Pacific Northwest Laboratory
Richland, Washington, U.S.A.

The radioactive waste isolation safety assessment program (WISAP) is being performed for the United State Department of Energy. A large bank of data have been collected under this program. A part of the program is devoted to the evaluation of radionuclide-geomedia interactions in order to determine the behavior of different elements in the environment. This paper introduces and briefly summarizes data collected thus far on methods of determining adsorption coefficients, thermodynamic predictions of stable solid and solution species, Kd data, and studies for determining adsorption mechanisms.

INTRODUCTION

Preliminary analysis has shown that by using geologic selection criteria [1] for selecting a repository site, the probability of a repository breach and resulting hazard to man is very low. In the event that the repository is breached by intruding groundwater, radioactive wastes will be leached and radionuclides brought into solution. The most important barrier to radionuclide movement, aside from engineered barriers, is retardation due to interaction between radionuclides and geologic media (Figure 1). Because of higher than normal temperatures and the radiation field in the repository, groundwater leaving the disposal site is expected to contain radionuclides. The final radionuclide concentrations in any groundwater that may come in contact with the biosphere are of major concern.

For full comprehension of radionuclide behavior, one must understand the radionuclide chemistry as it affects various retardation mechanisms. Figure 1 lists the types of data required: 1) accurate thermodynamic data on solid phases and solution species that may be present or may form in the geomedia, 2) adsorption-desorption isotherms, 3) equations that quantify effects of different factors as they influence exchange, adsorption, precipitation and dissolution, and solution species, and 4) mineralogical and physico-chemical properties of the geologic media. All such data are not presently available. Because of time and money constraints, thorough investigation of the different radionuclide interaction mechanisms is not possible. Therefore, an empirical approach in combination with a limited mechanistic approach was chosen by the Waste Isolation Safety Assessment Program (WISAP) to evaluate the radionuclide geologic media interactions. In the empirical approach, the influence of exchange and surface adsorption and of precipitation and dissolution are evaluated by determining adsorption distributions (Kd or more appropriately retardation factor) in the laboratory for a large number of well characterized geomedia and solution types. The objectives of this paper are to discuss different methods of determining Kd and the progress on accumulating a data bank for geomedia-radionuclide interactions.

METHODS OF DETERMING Kd

The methods of determining Kd can be grouped into two categories, the static or "batch" technique and the dynamic or column methods. In the static method, solution traced with radionuclides is shaken with samples of geomedia for a period before sampling and analysis. Investigation of dynamic systems entails the study of a tracer solution as it flows through geologic media. The dynamic methods (low pressure flow, axial filtration, high pressure intact core) differ in the type of pressure required to force liquid through the geomedia. The advantages and disadvantages of different methods of Kd determination are discussed by Relyea, et al. [2]. The two methods (static and dynamic) for radionuclide-geologic media interaction studies are complementary, with the static method being most useful for screening investigations of radionuclide behavior in a variety of systems and for estimating the time needed to attain equilibrium. Static radionuclide adsorption distributions may then be compared to retardation factors obtained from dynamic systems under similar conditions to verify results. Crushed or uncrushed material may be used in either method. Solutions similar in composition to the final solutions in static tests may also be used in dynamic experiments. The most pressing current need is for a relationship to relate Kd values obtained on crushed material or uncrushed pieces to the natural geomedia that may contain variable amounts of tortuous fissures and cracks.

Since a large number of factors inherent in the geomedia and solution (Figure 1) can affect the Kd, a small change in procedures may give different results. Therefore, thorough characterization of solution and geomedia should be done for all Kd determinations. A study is currently under way to evaluate the Kd determination methods.

The Kd values for Cs and Sr determined by the batch, flow through column (crushed material), and axial filtration methods for simple mineral-solution systems show favorable agreement [3]. Channel chromatographic and batch Kd

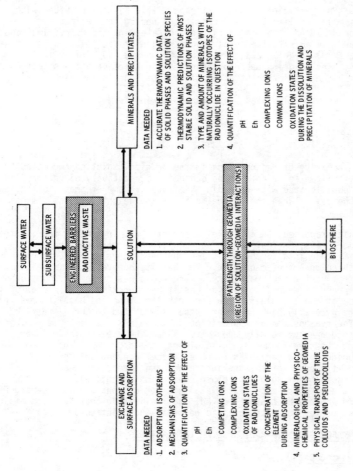

Figure 1   Factors that determine the concentration of radionuclides in solutions leached through the waste.

techniques used to determine the adsorption capacities of shale for Sr and Ni from dilute calcium nitrate solutions also compare favorably [4]. Although it is too premature to make definite conclusions, the different experimental methods give comparable results on crushed media for simple systems. We plan to relate sorption onto fissures versus crushed material via normalization of surface area. At present we have no conclusive data to verify this approach. More complex systems (natural rocks as opposed to pure minerals, natural groundwater as opposed to synthetic salt solutions, and multivalent nuclides such as the Pu) should and will also be studied by various experimental techniques.

In addition, an interlaboratory comparison [5] of batch Kd values of oolitic limestone (Bedford, Indiana) and basalt (Sentinel Gap, Washington) for Cs, Sr, and Pu was also performed in order to find the precision of determining Kd by the batch method. The participating laboratories (Table 1) were provided with homogeneous samples of limestone and basalt and were given recipes for groundwater preparation. The solution used to determine the Kd of limestone was between 8.0 and 8.4 pH with the total concentration of various ions in ppm as Ca^{2+} = 50, Mg^{2+} = 3.6, Na^+ = 2.3, K^+ = 0.4, HCO_3^- = 153, SO_4^{2-} = 14.4, and Cl^- = 3.5. The solution used to determine the Kd of basalt was between 7.7 and 8.2 pH with the total concentration of various ions in ppm as Ca^{2+} = 6.5, Mg^{2+} = 1.0, Na^+ = 30, K^+ = 9, HCO_3^- = 58, SO_4^{2-} = 23, Cl^- = 16, and F^- = 0.7. The Kd was calculated using the following formula:

$$Kd = \left[\frac{C_0 - C}{C}\right]\left[\frac{V}{W}\right]$$

in which Co = activity of radionuclide in the original solution corrected for loss of activity from solution with no geomedia, C = activity of radionuclide in the solution equilibrated with geomedia, V = total volume of the solution, and W = weight of the soil. When the volume is in ml and weight in g the dimensions of Kd are ml/g.

The results reported [5] in Table I indicate that in general for all the isotopes compared (Cs, Sr, Pu), there was considerably more disagreement in Kd values for limestone than for basalt. The observed Kd values of Cs for limestone ranged over three orders of magnitude. For basalt, however, the Kd values of Cs varied by a factor of approximately two. The Kd values of Sr for limestone varied by an order of magnitude and for basalt by approximately a factor of two. A quick glance at the Kd values of Pu for both limestone and basalt shows that the values are generally high and that these vary by two to three orders of magnitude. The reasons for poor precision in Kd values (Table 1) determined by the batch method are not readily apparent. Since the data in Table 1 was obtained using the same geomedia and solutions, the poor precision may be due to some of the uncontrolled variables (such as tracer concentrations, Eh, preparation of tracer solution, shaking method and speed, temperature, variation of pH of the equilibrium solution, phase separation technique). This study reemphasized the need for mechanistic studies, strict control and understanding of factors and procedures that affect Kd, and realization that Kd is an empirical parameter most useful for site specific geomedia and solutions.

DATA BANK FOR GEOMEDIA-RADIONUCLIDE INTERACTIONS

Thermodynamic Data

Although, at the present, meaningful data for interaction of radionuclides with all geomedia are not available [6], such data are accumulating rapidly 7,8. In order to best utilize the data generated by many researchers from the United States and abroad, we are using chemical computer models and computerized data files. The data files of existing thermodynamic equilibrium computer models [9,10,11,12] are being updated to include actinide and long lived fission products. The data files consist of standard Gibbs free energies and entropies of formation for a wide variety of minerals, aqueous species and gases [13]. From these data the distribution of an element between solid, liquid and gaseous species may be predicted assuming equilibrium conditions. Based upon the

Table I. Interlaboratory Comparison of Kd (ml/g) Values for Limestone and Basalt for Cs, Sr and Pu* [5]

| Cs | | | | Sr | | | | Pu | | | | Participating Laboratories |
| Limestone | | Basalt | | Limestone | | Basalt | | Limestone | | Basalt | | |
pH	Kd	pH	Kd	pH	Kd	pH	Kd	pH	Kd**	pH	Kd**	
7.9	880±160	8.7	380±70	7.9	14.9±4.6	8.7	92±3	7.9	1090±120[a] 690±320[b]	8.6	200±170[a] 15±2[b]	PNL
7.9	65±2	7.8	401±21	8.0	5.4±0.3	8.1	68±17	7.6	1027±41[a] 2134±77[c]	7.5	8.8±3.4[a] 67±14[c]	ANL
8.5	88±1	8.4	265±4	8.5	1.4±0.2	8.4	81±1	8.1	4745±4931[a] 3198±512[c]	8.0	1078±716[a] 843±658[c]	LASL
ND	60±30	ND	290±70	ND	2.7±0.5	ND	45±1	ND	770±36[b]	ND	580±12[b]	LLL
8.2	227±14	8.3	380±5	8.3	5.9±0.2	8.4	89±5	8.2	198±50[a]	8.2	592±134[a]	ORNL-I
ND	663±61	ND	453±12	ND	9.3±2.4	ND	93±6	ND	63000±19000	ND	312±66[a]	ORNL-II
ND	6.8±0.6	ND	255±7	ND	13.4±0.6	ND	73±4	ND	2616±1135[b]	ND	89±17[b]	RHO
7.8	1.3±0.4	7.4	31±2	7.6	1.8±0.5	7.4	41±6	7.5	547±356[b]	7.5	1000±590[b]	AEC (Canada)
ND	48.9±5	ND	296±10	ND	2.4±0.1	ND	55±2	ND	1009±162[d]	ND	221±147[d]	LBL

* The limestone and basalt samples were crushed to a 297 to 840 m size. All Kd's are for a 297 to 840 m size. All Kd's are for a 7 day equilibration period except ORNL-II which are for a 29 day equilibration period. The ND in the data means "not determined".

** Equilibration solution treatment, a = centrifuged at approx. 4000 g and filtered through 0.2 to 0.45 m filter, b = centrifuged at approx. 4000 g, c = absorbed activity was determined directly from the geomedia sample instead of calculating from the change in concentration of the radionuclide in solution, d = based on original influent which was not corrected for the loss of activity from the solution not containing geomedia.

- 67 -

thermodynamic data, a summary of stable solution species that may be present in relatively oxidizing and reducing environments is given in Table II.

Kd Data Bank

The second type of computerized data file, the Kd data bank, consists of a tabulation of nuclide distribution coefficients along with a complete information about the geomedia and solution [14]. To properly assess whether a nuclide distribution coefficient (Kd) or retardation factor previously determined is applicable to a scenario under study, it is necessary to have a complete description of the geologic media, contacting solution, nuclide properties and experimental details. Too often in available literature Kd values are presented without these necessary details. Without the supporting information it is difficult to assign confidence limits on the appropriateness of the data. The Kd data generated in the last two years for the WISAP program are currently being installed on a Pacific Northwest Laboratory (PNL) information-retrieval system. Data generated under the WISAP Program [7,8] and other literature data will also be incorporated into the data bank at a future date. This capability will allow to sort the data automatically and to search Kd data by computer using details regarding the author, experimental conditions, type of geologic material and its properties, solution properties, and information about species, concentration, and nature of radionuclide. For example, if a listing of all Kd data for basalt is needed, one can ask the computer to tabulate all basalt data. If one wants to compare Kd's from a salt brine across all rock types one can sort on brine, etc. The computer data bank sorting capability will also be linked to statistical programs such that selected data will automatically be transferred from the data bank to the statistical program and after manipulation the results will be displayed on a screen or produced as hard copies. In this fashion one can get means, standard deviations, and empirical predictor equations for Kd as a function of minerals, surface area, cation-exchange capacity, groundwater composition, etc.

The data bank will be set up such that the operator is lead through the procedure step by step. It will be usable by nongeochemists. We want the data information-retrieval system to be usable by modelers without necessitating a lot of technical interpretation. If a modeler needs Kd's for a particular scenario the information-retrieval system will inquire about the type of scenario and after the operator inputs on the keyboard, the computer will select possible choices and display them along with any other basic details requested.

Statistical Technique to Synthetize Kd Data

The statistical analysis and synthesis of the experimental data is accomplished using the adaptive learning "self learning" networks developed by Adaptronics, Inc. [15,16,17,18]. The Adaptive Learning Network (ALN) is a mathematical method, made possible by high speed computers, which fits a multinomial (a polynomial in many variables) to the multidimensional surface describing a radionuclide Kd. Thus, the ALN methodology is a powerful tool for use in data modeling instances where little or no knowledge exists regarding the functional relationship of dependent to independent variables. In order to construct a self-learning network, a list of important variables along with data representative of variety of situations that can occur in the system is needed. The next step is to construct a generalized equation to link an output value to each possible pair of input variables. Variables are initially taken two at a time in quadratic form to describe the Kd. Special purpose computer programs are used to find the coefficients (the weights assigned to the variables in the quadratic equation for the Kd) for each equation that make it best fit the data. These equations and variables that produce the smallest prediction errors are determined and retained. In the next step, quadratic equations are taken as new variables and the above procedure is repeated to obtain polynomial equations (up to eighth order) which predict the Kd. This process is repeated as additional layers of equations are added to the network and retained if they can improve its predicting ability. The training procedure is terminated when the model would become so adept at fitting the data used to train it that it would be unable to generalize to data not previously seen. Special algorithms are used to detect and avoid this condition. To make certain that the model has indeed discovered

Table II. Predominant Solution Species of Elements in a pH 4 to 9, pO_2 0.68 to 80, pCO_2 1.52 to 3.52, $pCl^- = pNO_3 = pSO_4 = 3.0$, $pF^- = 4.5$ and $pH_2PO_4 = 5.0$ Environment (Activities of Solution Ions in moles/l) Without Organic Ligands [6]

Elements	Little Affected by Oxidation-Reduction	In An Oxidizing Environment	In A Reducing Environment
		Predominant Solution Species of Elements	
Am	Am^{3+}, $AmSO_4^+$, $Am(OH)^{2+}$		
Sb		$HSbO_2^0$, $Sb(OH)_3^0$, $SbOF^0$, $Sb(OH)_4^-$	SbO^+
Ce	Ce^{3+}, $CeSO_4^+$		
Cs	Cs^+		
Co	Co^{2+}, $Co(OH)_2^0$		
Cm	Cm^{3+}, $CmOH^{2+}$, $Cm(OH)_2^+$		
Eu	Eu^{3+}, $EuSO_4^+$, $Eu_2P_2O_7^{2+}$		
I	I^-, IO_3^-		
Np		NpO_2^+, $NpO_2HPO_4^-$, NpO_2HCO_3	$NpOH^{3+}$, Np^{4+}
Pu		PuO_2^{2+}, $PuO_2(CO_3)(OH)_2^{2-}$, PuO_2^+	$PuOH^{2+}$, Pu^{3+}
Pm	Pm^{3+}		
Ra	Ra^{2+}		
Ru		$Ru(OH)_2^{2+}$, RuO_4^-, RuO_4^{2-}	RuO_4^-
Sr	Sr^{2+}		
Tc		TcO_4^-	Tc^{2+}
Th	ThF^{3+}, $Th(OH)_3^+$		
$3H$	H^+, $3H$-O-H		
U		UO_2^{2+}, UO_2F^+, $UO_2(OH)_2^0$, $UO_2(CO_3)_3^{4-}$	UO_2^{2+}, UOH^{3+}, UO_2^+, $UO_2(CO_3)_3^{4-}$
Zr	$Zr(OH)_4^0$, $Zr(OH)_5^-$, ZrF^{3+}		

for itself the pertinent physical laws, additional experimental data not used in the training phase are introduced to test the ability of the model to generalize on its prior experience in dealing with new situations.

The performance of the ALN method was compared with the regression method [15] by using the Kd data generated by Relyea et al. [19]. In this evaluation, Kd as the dependent variable was related to different properties of the geomedia (cation exchange capacity, surface area), solution properties (Na^+, Ca^{2+}, Cl^-, HCO_3), time (time dependent Kd experiments), and pH and Eh of the mineral-solution suspensions. A part of the results are shown in Table III. The results [15] presented in Table III show that there was a considerable inprovement in predicting Kd when the ALN method was used than when the regression method was used. While performing clustering techniques on these data [19], it was discovered that the adsorption coefficients for silicate minerals were a function of the structural group (phyllosilicates, tektosilicates, and inosilicates). Although studies are under way, it is not known at present whether the rock Kd can be modeled from the information regarding its chemical, physical, and mineralogical properties. We plan to use all available data to generate equations that would relate Kd as a function of geologic media, groundwater and nuclide characteristics. Although the derived relationships will not prove "cause and effect" or confirm sorption mechanisms, they will allow prediction of trends. Thus from Kd data on a finite number of rock and water types, estimates of Kd for a nuclide for other rock-water environments not directly studied can be made if precautions, such as avoiding extrapolation beyond the ranges of variables used to construct the empirical mathematical relationship, are observed.

Adsorption Data Based Upon Mechanistic Studies

The Kd concept, computer data bank, and empirical Kd predictor equations do not rigourously lend themselves to an understanding of controlling mechanisms. A limited number of mechanistic studies are under way at present, with more emphasis to be placed upon this type of study in following years. Studies conducted thus far at PNL have revealed that:

1) in radioactive waste disposal cribs containing discrete PuO_2 particles, the concentration of soluble Pu is predictable from a solubility relationship [20],

2) the solubility of PuO_2 and Pu(IV) hydroxide actually observed under environmental conditions differ from theoretical predictions [20,21],

3) Pu(IV) polymer formation can be predicted from pH and total soluble Pu concentration [22],

4) Pu(V) rather than Pu(VI) may be the dominant species in environmental samples in relatively oxidizing conditions [23], and

Microautoradiography studies of rock thin sections performed at Lawrence Berkely Lab (LBL) and Los Alamos Scientific Laboratory (LASL) have shown that the active sorption sites in bulk rocks (basalt, quartz monzonite, argillite, tuffaceous alluvium and tuff) for actinides are generally the secondary mineral alteration products such as clays, zeolites etc. [24,25].

Accelerated weathering studies on basalt and argillite increased the sorption tendencies for most isotopes studied (Cs, Sr, Pu, Am, Np). Weathering of granite showed no observable changes in sorption [26].

Systematic ion-exchange studies versus supporting electrolyte concentration at constant pH show the adsorption of Sr and Cs is approximately as expected from conventional idealized ion exchange equations. The trivalent ions Eu and Am however do not conform to idealized equations. Details on this work performed at Oak Ridge National Laboratory (ORNL) can be found in these [3,8] references.

Table III. Performance of ALN and Regression (Reg) Techniques On Data Not Used In Model Synthesis [15]

Method	Tc (0 to 22) [1]		Sr (0 to 1680) [1]		Cs (1.3 to 140000) [1]		Np (0 to 7324) [1]	
	Average Absolute Error	Correlation Coefficient	Average Absolute Error	Correlation Coefficient	Average Absolute Error	Correlation Coefficient	Average Absolute Error	Correlation Coefficient
Time Independent								
ALN	1.3	.88	75.4	.91	5481	.67	15.4	.87
Reg.	2.4	.62	91.0	.85	6330	.52	25.4	.31
Time Dependent								
ALN	1.3	.65	59.4	.96	4151	.41	18.4	.74
Reg.	1.5	.60	79.4	.90	4452	.18	23.3	.35

1/ Numbers in parentheses denote the range of values of K_d for the particular nuclide.

In other studies at ORNL [4] it was found that redox sensitive elements Np and Tc should not be stable in their mobile NpO_2^+ and TcO_4^- forms in contact with igneous rock-anoxic groundwater suspensions. Np and Tc added in thier mobile oxidized states are reduced to much less mobile species by these environments. The current safety assessments which consider the oxidized forms (NpO_2^+ and TcO_4^-) only may be overestimating their potential hazard to the public.

At Argonne National Laboratory (ANL) kinetic and partioning data for Am in fissures of unpermeable rock were used as input to a transport model which includes time dependent sorption (kinetic effects). Laboratory infiltration experiments into fissures were compared with model predicted behavior. The faborable comparison emphasizes the need to consider reaction kinetics in the prediction of nuclide migration [28].

BIBLIOGRAPHIC REFERENCES

1. Brunton, G. D. and W. C. McClain: "Geologic Criteria for Radioactive Waste Repositories", Y/OWI/TM-47 (1977).

2. Relyea, J. F., Dhanpat Rai, and R. J. Serne: "Interaction of Waste Radionuclides with Geomedia: Program, Approach, and Progress", U.S. Dept. of Energy Rep. PNL-SA-7289 (1978).

3. Meyer, R. E., S. Y. Shiao, P. Rafferty, J. S. Johnson, Jr., I. L. Thomas, and K. A. Kraus: "Systematic Study of Nuclide Sorption on Select Gelogic Media", Waste Isolation Safety Assessment Program Task 4 Contractor Information Meeting Proc. pp. 343-369, U.S. Dept. Energy Rep. PNL-SA-6957 (1977).

4. Francis, C. W., M. Reeves III, R. S. Fisher, and B. A. Smith: "Soil Chromatograph K_d values", Waste Isolation Safety Assessment Program Task 4 Contractor Information Meeting Proc. pp. 403-431, U.S. Dept. Energy Rep. PNL-SA-6957 (1977).

5. Relyea, J. F. and R. J. Serne: "Controlled Sample Program; Interlaboratory Comparison of Batch Kd's", U.S. Dept. of Energy Rep. PNL-2872 (1979).

6. Ames, L. L. and Dhanpat Rai: "Radionuclide Interactions With Soil and Rock Media", U.S. Environmental Protection Agency Rep. EPA 520/6-78-007.

7. Serne, R. J. (Chairman): "Waste Isolation Safety Assessment Program Task 4 Contractor Information Meeting Proc.", U.S. Dept. of Energy Rep. PNL-SA-6957 (1977).

8. Serne, R. J. (Chairman): "Waste Isolation Safety Assessment Program Task 4 Contractor Information Meeting Proc.", U.S. Dept. of Energy Rep. PNL-SA-7352 (1979).

9. Helgeson, H. C.: "Evaluation of Irreversible Reactions in Geochemical Processes Involving Minerals and Aqueous Solutions-I Thermodynamic Reactions", Geochim. Cosmochim, Acta. 32, 853-877 (1968).

10. Helgeson, H. C., R. M. Garrels, and F. T. Mackenzie: "Evaluation of Irreversible Reactions in Geochemical Processes Involving Minerals and Aqueous Solutions-II Applications", Geochim Cosmochim. Acta, 32, 455-482 (1968).

11. Helgeson, H. C., T. H. Brown, A. Nigrini, and T. A. Jones: "Calculations of Mass Transfer in Geochemical Processes Involving Aqueous Solutions", Geochim Cosmochim. Acta, 34, 569-592 (1970).

12. Helgeson, H. C.: "Kinetics of Mass Transfer Among Silicates and Aqueous Solutions", Geochim. Cosmochim. Acta, 35, 421-469 (1971).

13. Apps, J. A., L. V. Benson, J. Lucas, A. K. Mathur, and L. Tsao: "Theoretical and Experimental Evaluation of Waste Transport in Selected Rocks", Waste Isolation Safety Assessment Program Task 4 Contractor Information Meeting Proceedings, pp. 189-309, U.S. Dept. of Energy Rep. PNL-SA-6957 (1977).

14. Burkholder, H. C., J. Greenborg, J. A. Stottlemyre, D. H. Bradley, J. R. Raymond, and R. J. Serne: "WISAP Technical Progress Report for FY 1977", U.S. Dept. of Energy Rep. PNL-2642 (1978).

15. Mucciardi, A. N., and E. C. Orr: "Statistical Investigation of the Mechanics Controlling Radionuclide Sorption", Waste Isolation Safety Assessment Program Task 4 Contractor Information Meeting Proc., pp. 151-188, U.S. Dept. of Energy Rep. PNL-SA-6957 (1977).

16. Barron, R. L.: "Theory and Application of Cybernetic Systems: an Overview", Proc. IEEE 1974 National Aerospace Electronics Conference (NAECON '74), pp. 107-118 (1974).

17. Mucciardi, A. N. and E. E. Gose: "An Automatic Clustering Algorithm and Its Properties in High-Dimensional Spaces", IEEE Trans. Computers, SMC-2, pp. 247-254 (1972).

18. Mucciardi, A. N.: "Elements of Learning Control Systems with Applications to Industrial Processes", Proc. 1972 IEEE Conference on Decision and Control, pp. 320-325 (1972).

19. Relyea, J. F. R. J. Serne, Dhanpat Rai, and M. J. Mason: "Batch Kd Experiments With Common Minerals and Representative Groundwaters", Waste Isolation Safety Assessment Program Task 4 Contractor Information Meeting Proceedings, pp. 125-150, U.S. Dept. of Energy Rep. PNL-SA-6957 (1977).

20. Rai, Dhanpat, R. J. Serne, and D. A. Moore: "Identification of Plutonium Compounds and Their Solubility in Soils", Waste Isolation Safety Assessment Program Task 4 Contractor Information Meeting Proceedings, pp. 107-124, U.S. Dept. of Energy Rep. PNL-SA-6957 (1977).

21. Rai, Dhanpat and R. J. Serne: "Solid Phases and Solution Species of Different Elements in Geologic Environments", U.S. Dept. of Energy Rep. PNL-2651 (1978).

22. Rai, Dhanpat and R. J. Serne: "Solution Species of ^{239}Pu in Oxidizing Environments: Polymeric Pu IV ", U.S. Dept. of Energy Rep. PNL-SA-6994 (1978).

23. Rai, Dhanpat, R. J. Serne, and J. L. Swanson: "Solution Species of ^{239}Pu V in the Environment", U.S. Dept. of Energy Rep. PNL-SA-7027 (1978).

24. Silva, R. J. L. V. Benson, and J. A. Apps: "WISAP 4: Collection and Generation of Transport Data", Waste Isolation Safety Assessment Program Task 4 Contractor Information Meeting Proceedings, U.S. Dept. of Energy Rep. PNL-SA-7352 (to be issued in 1979).

25. Thompson, J. L. and K. Wolfsberg: "Applicability of Microautoradiography to Sorption Studies", Waste Isolation Safety Assessment Program Task 4 Contractor Information Meeting Proceedings, U.S. Dept. of Energy Rep. PNL-SA-7352 (to is issued in 1979).

26. Barney, G. S. and P. D. Anderson: "The Kinetics and Reversibility of Radionuclide Sorption Reactions with Rocks", U.S. Dept. of Energy Rep. RHO-ST-19 (1978).

27. Seitz, M. G., P. G. Rickert, S. M. Fried, A. M. Friedman, and M. J. Steindler: "Transport Properties of Nuclear Waste in Geologic Media", Waste Isolation Safety Assessment Program Task 4 Contractor Information Meeting Proceedings, U.S. Dept. of Energy Rep. PNL-SA-7352 (to be issued in 1979).

Discussion

G. LENZI, Italy

What is the influence of the parameter "time" on the equilibrium between solid and liquid phases ? Have you considered this variable ?

D. RAI, United States

Yes, we have considered this variable. In our analysis we determine the concentration of radionuclides in the equilibrating solution at three different time intervals (approximately 7 days, 14 days, and 28 days) to make certain that an apparent equilibrium has been reached.

G. LENZI, Italy

What is the importance, in your opinion, of the solution temperature ?

D. RAI, United States

It may be important to investigate the effect of temperature for two different reasons. First, it may help to understand the adsorption mechanisms ; second, the temperatures in a waste repository are expected to be greater than at the surface. Most experiments conducted so far have been at 25°C. However, studies are underway to determine adsorption coefficients at temperatures which are expected in waste repositories.

F. GIRARDI, CEC

You have shown data obtained by intercomparison which show large variations. How do you make a choice among them ? You have indicated that these data will be used in a computer programme to predict radionuclides migration in the geosphere, if I understood correctly, and I wonder how a reasonable prediction can be done on the basis of such data.

D. RAI, United States

In general, there was a fair amount of agreement between Kd values obtained for Sr and Cs in basalt. In our data base we rate the input data according to their accuracy. Based upon such evaluations we can then put a confidence limit on the predicted Kd for a particular scenario we may be interested in. As the quality of the input data increases, the Kd values for a particular scenario would also improve.

P. BO, Denmark

Has any direct correlation between Kd measured and method of separating liquid and solid phases (centrifugation, filtration) been established ?

D. RAI, United States

We have made a few observations only ; but no definite conclusions has been reached. In general, for the highly adsorbing elements, the Kd increases when the degree of separation between solid and solution phases is increased by any means (ultra filtration or a combination of supercentrifugation and filtration).

D.L. RANCON, France

Nous confirmons que la grande dispersion dans les mesures de Kd, toutes les autres conditions étant égales, peut être due au processus expérimental, par exemple à la séparation des phases par centrifugation ou par filtration ; ceci est particulièrement sensible dans le cas du Pu, de l'Am ou des terres rares. Existe-t-il des recommandations pour une standardisation de ces mesures expérimentales de Kd ?

D. RAI, United States

I am glad to know that other researchers have also found that small changes in experimental methods can cause large variations in measured Kd. Our study on comparing different methods of Kd determination is in progress. At the end of this study, we will make recommendations regarding the experimental set up of suitable procedure or procedures.

L.H. BAETSLE, Belgium

I would like to make the following comments :

- Kd's determined in equilibrium conditions and under pure ion exchange mechanism should give the same results, from one laboratory to another, if the solution equilibrated with the solid phase is the same.

- Standardization of Kd is only valuable for the purpose of screaning geological formations.

- The true Kd's should be determined in groundwater conditions.

- If precipitates or colloids are present in the "trace" solution, the concept "Kd" is meaningless.

D. RAI, United States

I agree with the comments in general. However, I was pointing out that the equilibrium conditions varied somewhat from one lab to another and that the methods used to separate the solid phase from the solution (the degree of separation) may have a marked effect on the calculated Kd.

F. GIRARDI, CEC

I want to support strongly the statement of Mr. Baetsle. Standardizing a Kd measurement procedure on a simply empirical basis may be rather dangerous when mechanisms other than ion exchange are intervening.

D. RAI, United States

We realize that Kd's, as presently used, are empirical.
During a Kd determination, we measure all the factors that may
affect it. Eventually, it may be possible to relate the Kd to some
of these factors. Regardless of whether a Kd is empirically determined or based upon mechanistic studies, anything we can do to
obtain precise values, in our measurements, would help.

P. BO, Denmark

Standardization of Kd measurements can only be used for
inter laboratory comparisons, but does not normally reflect the
conditions during transport in geological formations especially
when the question of mobility of colloids is involved.

D. RAI, United States

In our measurements, we make an effort to assure that
precipitation and colloid formation (in the stock solutions used
to measure Kd) is minimal. Thus the results obtained should be a
function of the nature of geologic media as well as of other experimental and environmental conditions.

ESTIMATION DES VITESSES DES ECOULEMENTS SOUTERRAINS REGIONAUX EN MILIEU CRISTALLIN AU MOYEN D'UN MODELE DE SIMULATION

M. Bonnet, M.-L. Noyer, P. Vaubourg
Bureau de Recherches Géologiques et Minières
Département Hydrogéologie
Orléans (France)

Résumé

Les travaux de recherche que nous présentons ici ont été réalisés sous contrat à frais partagés entre la Commission des Communautés Européennes et le Commissariat à l'Energie Atomique, dans le cadre du programme d'action indirecte "Gestion et stockage des déchets radioactifs".

La connaissance des écoulements souterrains est importante pour juger du confinement du stockage en milieu cristallin. Dans ce but, un modèle de simulation a été construit pour estimer les écoulements souterrains et apprécier les vitesses de transfert en profondeur (intensité et direction) à des cotes correspondant à celles où sont envisagés les sites de stockage. Ce modèle a été éprouvé quant aux conditions aux limites de surface aux conditions rencontrées dans un bassin versant expérimental dont nous esquisserons une brève description avant de présenter les principaux résultats du modèle.

o ° o

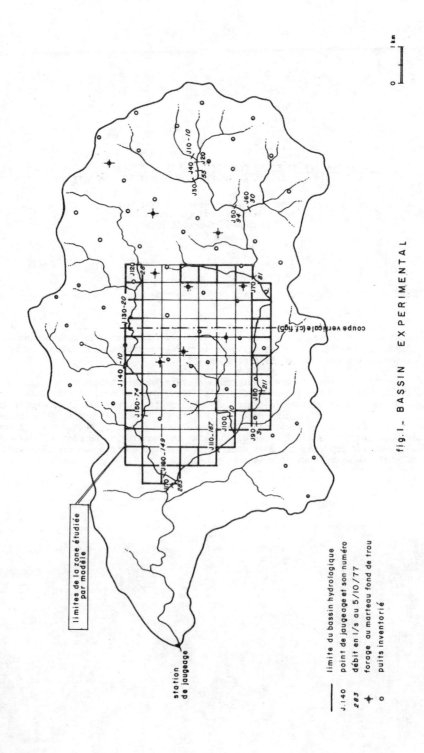

fig. I.- BASSIN EXPERIMENTAL

limites de la zone étudiée
par modèle

limite du bassin hydrologique
J.140   point de jaugeage et son numéro
283   débit en l/s au 5/10/77
✦   forage au marteau fond de trou
○   puits inventorié

station
de jaugeage

1. CADRE GEOLOGIQUE ET STRUCTURAL DU BASSIN EXPERIMENTAL

Le bassin expérimental d'une surface de 81 km^2 est caractérisé par la présence de granite à biotite seule et à biotite et cordiérite. Le faciès à biotite seule est dominant.

Ces granites, d'âge tardicadomien, présentent une altération classique en boules s'accompagnant d'une arénisation d'intensité variable. L'épaisseur des arènes varie de quelques mètres à 30 m. Des filons de quartz de direction N-S, intrusifs dans les granites, sont fréquents.

Du point de vue structural, les études de terrain et l'observation des photographies aériennes ont permis de mettre en évidence plusieurs directions de fracture qui sont par ordre d'importance : N 150° E, N 30°, N 80°, N 110° E et NS. Ces fractures sont subverticales.

Les fractures de direction N 150° E sont les plus nombreuses et les plus continues. Elles semblent se disposer en relais et constituer des couloirs de forte densité de fracturation. L'un des couloirs partage le bassin versant en deux zones d'importance sensiblement égale et à densité de fracturation moyenne.

2. DISPOSITIF DE MESURES

Les écoulements de surface sont parfaitement contrôlés par une station permanente. Des mesures complémentaires ont été réalisées en octobre 1977, période très proche de l'étiage.

De son côté, le milieu fissuré a été reconnu par une série de forages de 75 m de profondeur sur lesquels des slugs tests ont permis une première approche des valeurs de perméabilité. De même des pompages d'essai d'assez longue durée réalisés dans une zone voisine présentant des caractéristiques géologiques comparables à celles du bassin expérimental fournissent des indications intéressantes sur les perméabilités du milieu fissuré entre 0 et 100 m de profondeur.

3. DESCRIPTION DU MODELE DE L'ECOULEMENT SOUTERRAIN

La simulation numérique porte sur les écoulements se manifestant dans un "bloc diagramme" s'appuyant sur des limites naturelles constituées par des rivières (fig. 1). Ce bloc est délimité par : des parois verticales, un fond horizontal à -990 m et la surface du sol. Les rivières sont les seuls exutoires du système.

3.1. Hypothèses et conditions aux limites

Les hypothèses suivantes ont été retenues :

- les écoulements sont tridimensionnels et obéissent à la loi de DARCY. En régime permanent les débits obtenus sont donc proportionnels aux perméabilités imposées pour une répartition donnée de la charge hydraulique ;
- le milieu est continu et isotrope. Les zones de discontinuité telles que les fractures n'ont pas été représentées ;
- le milieu est "régulièrement" hétérogène ;
- la perméabilité à la surface étant K_S, on impose une loi de décroissance avec la profondeur z, donnée par $K(z) = K_S \, 10^{-z/500}$ (z en m). A la limite inférieure du domaine située à -1000 m la perméabilité est égale à $K_S/100$.

Les conditions aux limites suivantes ont été introduites :

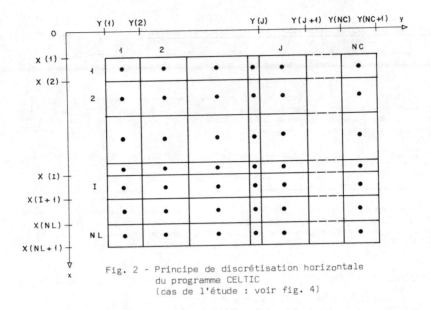

Fig. 2 - Principe de discrétisation horizontale
du programme CELTIC
(cas de l'étude : voir fig. 4)

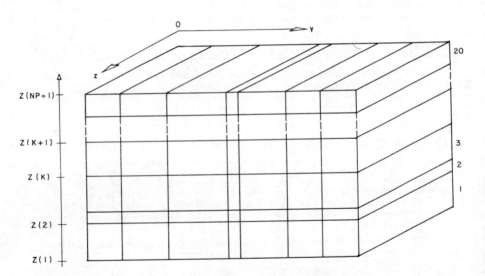

Fig. 3 - Vue perspective du parallélépipède simulé

NC : nombre de colonnes)
NL : nombre de lignes   )
NP : nombre de couches dans un plan vertical

- la piézométrie est imposée dans tout le domaine. Entre les cours d'eau elle est supposée suivre la surface topographique ;

- les limites verticales sont imperméables ainsi que la limite horizontale du fond.

3.2. Modèle utilisé et méthode de calcul

Le programme de calcul utilisé calcule les charges et les débits par une méthode aux différences finies classique (surrelaxation). Le domaine d'étude a donc été discrétisé en "couches", chaque couche étant elle-même redécoupée dans le plan horizontal. Les pas de discrétisation dans les trois directions ne sont pas nécessairement uniformes (fig. 2, 3 et 4).

Dans le cas présent, un découpage uniforme (mailles carrées de 500 m de côté (fig. 4) a été adopté dans le plan horizontal ; et verticalement, la discrétisation a été choisie en fonction de la topographie et de la loi de décroissance de la perméabilité. Le bloc étudié a donc été découpé en 20 couches (de 10 m à 300 m d'épaisseur) comportant 83 mailles utiles par couche, c'est-à-dire 1660 mailles utiles.

Enfin le programme permet de tracer des cartes piézométriques soit en plan horizontal, soit en coupe verticale. En adoptant une valeur de porosité on peut également tracer les lignes de courant et établir l'ordre de grandeur des temps de transferts.

4. RESULTATS

Les mesures de débits en période proche de l'étiage (emplacement et valeurs en l/s reportés sur la figure 1) ont permis d'évaluer le débit moyen d'étiage du bassin expérimental à 200 l/s environ. La partie du bassin versant concernant le bloc étudié est 0,4 à 0,5 fois la superficie de celui-ci, soit un débit correspondant de 80 à 100 l/s. En prenant comme hypothèse de calcul une perméabilité Ks de 1.10^{-5} m/s, et la loi de décroissance décrite ci-dessus, le débit global aux exutoires calculé est de 550 l/s. Les perméabilités étant proportionnelles aux débits, on est conduit à adopter une perméabilité Ks de 1,4 à $1,8.10^{-6}$ m/s, qui est du même ordre de grandeur que celle trouvée ($3,6.10^{-6}$ m/s) au cours des pompages d'essai.

En adoptant une porosité cinématique de 2.10^{-4}, on a déterminé pour chaque ligne de courant l'ordre de grandeur des transferts en jours (fig. 5).

Sous sa forme actuelle la conception du modèle reste simple mais des perfectionnements restent possibles par une meilleure représentation de la variation de la perméabilité en fonction de la profondeur, la prise en compte de l'anisotropie et enfin par une meilleure approche des porosités en essayant aussi de se caler en régime transitoire.

BIBLIOGRAPHIE

BERTRAND (L.), SAUTY (J.P.) .- Simulation mathématique tridimensionnelle par différences finies des écoulements autour d'une exploitation minière, 1975.

BONNET (M.) .- Méthodologie des modèles de simulation en hydrogéologie .- *Thèse : doct. état, Nancy, 1978.*

SAUTY (J.P.) .- Computer simulation of pollution front movment .- In : *Groundwater pollution in Europe, Proc. Conference in Reading (England), 1972. Port Washington, Water Information Center, 1974, p. 69-81.*

SAUTY (J.P.) .- Contribution à l'identification des paramètres de dispersion dans les aquifères par interprétation des expériences de traçage .- *Thèse : doct.-ing., Grenoble, 1977.*

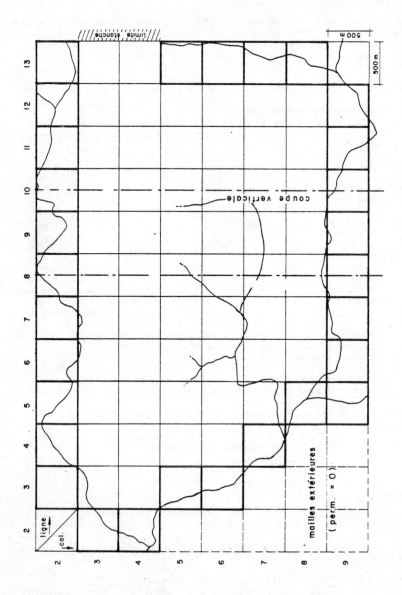

fig.4 - Discrétisation du domaine d'étude - coupe horizontale

- 84 -

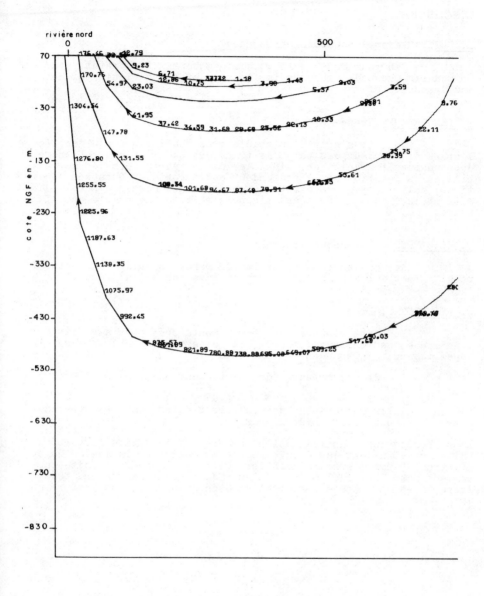

Fig. 5 - Lignes de courant et temps de transfert en jours
Extrait de la coupe verticale

Discussion

E. BÜTOW, Federal Republic of Germany

Can your model work with a geothermal temperature gradient to modifie the model results in function of the variation of density and viscosity of the water ?

P. PEAUDECERF, France

Non, il s'agit d'un modèle très schématique pour apprécier l'ordre de grandeur des temps de transfert que l'on peut attendre dans un massif granitique. Ainsi la loi des variations des perméabilités en fonction de la profondeur est théorique. Dans ces conditions, tenir compte des effets du gradient géothermique sur la densité et la viscosité de l'eau n'apporterait qu'une précision illusoire.

P. GLASBERGEN, The Netherlands

Are you sure that the law of Darcy is valid in fractured rock ? Great differences in transport time can be the result.

P. PEAUDECERF, France

En toute rigueur, les écoulements obéissent aux lois spécifiques des milieux fissurés, ce qui est observable aux voisinages des puits. Cependant, étant donné :

- l'ordre de grandeur des distances considérées (100 à 1.000 m),

- la méconnaissance des valeurs réelles des paramètres de conductivité hydraulique,

nous pensons que l'approximation dans ce cas du milieu fissuré en milieu continu dont les écoulements sont régis par la loi de Darcy, n'introduit pas d'erreurs notables. Cette hypothèse est faite couramment pour les modèles des études hydrogéologiques en aquifères fissurés et ceci donne satisfaction dans la plupart des cas.

Experimental Investigation of the Behaviour of Long-Living Radioisotopes in a Natural Water Clay System

P. De Regge, A. Bonne, D. Huys, R. Heremans
Studiecentrum voor Kernenergie
Centre d'étude de l'Energie Nucléaire
Mol (Belgium)

Considering the geological context of Belgium, the S.C.K./C.E.N. started in 1974 a study of the suitability of a clay layer for the burial of solidified radwaste. Research on the behaviour of radionuclides in a clay medium, with special interest for their valency states, ionic and polymeric forms, is carried on. Solubilisation and precipitation of plutonium in different waters has been investigated as a function of time and pH, starting from plutonium in solution and calcined plutonium oxide. The distribution of plutonium over its different valency states and other soluble forms has been studied. The adsorption of plutonium on clay as a function of plutonium concentration and pH is reported for samples taken at different depths throughout the clay layer. As the distribution coefficients tend to decrease at decreasing plutonium concentrations the low concentration range has been extensively investigated using a ^{238}Pu tracer. The effect on sorption capacity of the clay layer caused by the long term storage of high level radioactive waste, such as γ-radiation doses and temperature elevation has been studied. A limited number of experiments has shown that the retention of iodine in a natural water clay system due to adsorption of iodine is negligible.

1. INTRODUCTION

In 1974 the S.C.K./C.E.N. started the study of the suitability of a particular clay formation present in its underground (the Boom clay), as receptacle for the burial of radwaste.

Fieldwork carried out by the S.C.K./C.E.N. on its own site (reconnaissance, sampling, experiments) and subsequent analyses allow to get site specific parameters that can be applied to laboratory experiments and modeling in the frame of safety analysis and to the conceptual design of a potential waste repository in the Boom clay.

In the context of site specific research the S.C.K./C.E.N. investigated the problem of migration of radionuclides in the Boom clay formation and the sand formations on top of it. The present paper deals with the research on the behaviour of plutonium and in a minor extent of iodine, in a specific water-clay and water-sand system.

2. SITE SPECIFIC DATA

The Boom clay (Oligocene), occuring in the northeastern part of Belgium, is found beneath the nuclear site of Mol at depths between -160 m and -260 m. It is covered by glauconite bearing sands of Neogene age : sands of Voort, of Antwerp, of Dessel and of Diest. These sands compose together one semiconfined aquifer overlying the Boom clay.

In the $< 20 \mu$ fraction, which constitutes about 70 % of the Boom clay the following clay minerals are present (relative aboundances given between bracklets) : vermiculte (3/10), illite (2.5/10), smectite (2/10), interstratified illite-montmorillonite (1.5/10) and chlorite + interstratified chlorite and vermiculite (1/10). The cation-exchange capacity of these minerals is variable, ranging from 10 up to 150 milliequivalents per 100 g. Experiments on the Boom clay indicated a cation-exchange capacity for this rock of about 20 meq/100 g. Anion-exchange capacity of clay minerals is known, but is considerably lower than the cation-exchange capacity.

The water content of the Boom clay is ranging from 20 to 25 %, but the composition of the clay formation water is not yet known exactly. Nevertheless research carried out until now, revealed that the clay-water may be considered as an electrolytic solution with high chemical activity of SO_4^{--}, PO_4^{---}, Ca^{++}, Mg^{++} and Na^+. Research is continuing to improve characterisation of the clay formation water.

Research was started on the behaviour of radionuclides in the geologic media of interest by using distilled water and Anversian water (of the aquifer overlying the Boom clay) as carrier solutions. The composition of the Anversian water, sampled in a well neighbouring the nuclear site, is given in Table I. The choice of this Anversian water was to simulate a realistic geologic solution which could accidentially intrude from above into a potential waste repository within the clay formation.

Chemical, granulometrical and mineralogical analyses on samples, taken through the whole thickness of the clay formation, suggest a rhythmicity and some variation of characteristics within the clay formation. For this reason several samples of the Boom clay, taken at different depths, were studied.

The overlying Neogene sand formations are extremely rich in glauconite (up to more than 50 %). This silicate is considered as a mixed-layered mineral with mica- and clay species as components. Because of the presence of this important clay mineral fraction in the Neogene formations we may consider them also as a clay-water system. The cation-exchange capacity of glauconite is reported to reach 40 meq/100 g. For investigating the samples of these formations distilled water and the formation water (Anversian) itself were used as carrier solutions.

TABLE I : Analysis of Anversian Water   (in mg/l)
(sample locality : Mol)

Ca^{++}	3.03	Cl^-	6.40
Mg^{++}	3.20	CO_3^{--}	201
Na^+	55.3		
K^+	7.71	pH	8.35

TABLE II : Variation of plutonium concentration in ground water as a function of time. Concentrations in mg liter^{-1}

Days \ pH	1.88	2.33	3.03	7.95	8.42	8.79	8.63	8.63
0	776	388	155	77	38	7.8	3.9	0.077
2	761	378	129	33	15	3.8	1.9	0.032
7	739	373	133	26	11	2.8	1.3	0.038
10	749	365	92	22	8.9	1.7	0.83	0.026
30	763	366	94	23	8.4	1.2	0.64	0.014
45	750	347	89	22	8.5	0.9	0.48	0.018
370	677	327	54	10	8.5	0.3	–	–

TABLE III : Dissolution of PuO_2 in contact with ground water and dilute chloride and ammonia solutions as a function of time

	Ground water				Distilled water			
	+ HCl			+ NH₄OH	+ HCl			+ NH₄OH
pH	1	3	6	9	1	3	6	9
Days								
1	45	38	64	44	128	115	22	112
4	110	16	25	–	219	12	8	129
6	100	11	25	56	189	17	2	24
11	80	18	21	17	–	396	12	15
18	108	11	10	11	257	–	1	129
34	163	23	6	11	589	163	4	20
74	325	56	19	36	471	78	26	34
134	588	111	7	14	657	201	487	29
200	–	–	9	14	1124	1092	–	100
211	915	240	25	15	1064	–	41	187
322	1282	–	7	20	1537	1352	125	1085
361	1683	1506	81	42	1823	1585	995	869
368	1576	2719	68	–	2960	1411	1215	1988
390	1938	2904	93	48	2320	1340	–	–

3. SOLUBILITY STUDIES

3.1. Preparation of plutonium solutions in ground water (Anversian)

A plutonium nitrate solution was diluted with ground water in order to obtain a series of concentrations covering the range 0.1 mg ℓ^{-1} to 1 g ℓ^{-1}. Because the plutonium was dissolved in nitric acid the more concentrated solutions were slightly acid, the more dilute solutions approached the pH of the ground water itself at about 8.5. Precipitation of plutonium occurred in all solutions and the concentration variation was followed as a function of time by sampling the supernatant after centrifugation of the solution. The results are presented in Table II. As expected for solutions with a pH higher than 4 more than 80 % of the initial plutonium present precipitated even for the low concentrations. Important variations occur up to 10 days after the preparation, the solution remaining relatively stable after this period.

3.2. Determination of the ionic species of plutonium in ground water

A series of tests has been conducted to evaluate the distribution of plutonium in the ground water solutions among its different ionic species. A technique has been adapted using selective extraction of the different valencies as shown below :

a) extraction of Pu (IV) by TTA at pH 0.4 ;

b) extraction of Pu (III + IV) by TTA at pH 4.2 in 10 N ammonium-acetate ;

c) extraction of Pu (III + IV + VI) by TTA at pH 4.2 in 2 N ammoniumacetate ;

d) coprecipitation of Pu (III + IV + V) by LaF_3 ;

e) extraction of Pu (IV) polymer by dibutylphosphate.

The extractions a, b and c have been applied to the solutions of Pu in ground water. Solutions of a pH above 3 have to be acidified to remove carbonate ions complexing plutonium before extraction b) or c) can be applied.

The Pu^{3+} was found to represent between 0.25 and 2.5 % of the total Pu. The Pu^{4+} state ranged from 0.2 to 10.8 %. Less than one percent was found to be present as PuO_2^{2+}. As a result the bulk of the plutonium is not present as an ionic species Pu^{3+}, Pu^{4+} or PuO_2^{2+}. The plutonium could be present as a polymeric or complexed forms which are still to be identitied by suitable techniques.

3.3. Solubility of plutonium in natural waters

To obtain an estimate of the concentration range of interest with respect to the migration studies the solubility of plutonium in the natural environment of the site was investigated. Plutonium is assumed to be present as plutonium dioxide as the result of incineration processes. A preliminary series of tests was therefore conducted with PuO_2 fired at 600 °C to 700 °C. About 5 mg was contacted with ground water and distilled water which had been adjusted to pH 1, 3, 6 or 9 with hydrochloric acid or ammonia. The plutonium activity in the supernatant was followed as a function of contact time by sampling and alpha-counting. The results are given in Table III. Similar experiments are running with plutonium silicates calcined at 1100 °C contacted with different water types.

On the other hand 50 grams of clay (-224 m) have been shaken with 500 ml of distilled water for 24 hours. The resulting solution was filtered through a 2.2 µ filter and part of it was boiled for three hours. To both solutions plutonium tracer was added at three different concentration levels resulting in all cases in a precipitation of part of the plutonium. The final concentration in the supernatant sulution is given in Table IV.

The results are scattered particularly at neutral or basic pH ranges. This is probably due to the existence of colloidal plutonium particles in the solution which are not separated by centrifugation and do not settle down because of their heat dissipation due to alpha decay. Nevertheless the concentration range of interest for migration studies can be defined downwards from a few mg per liter to the allowable concentrations for natural waters.

4. ADSORPTION STUDIES

4.1. Method

For the estimation of the sorption of Pu and I on Boom clay and glauconite sands the static batch technique was used. This technique allows to obtain easily a commonly used parameter to qualify the sorption (K_d) and the equilibrium concentration.

The sorption-desorption reaction between the samples and the carrier solutions was carried out by dropping 0.2 g of a prepared sample in a recipient filled with 10 ml of carrier solution. Good phase contact was ensured by placing the recipients in a rotating drum (about 29 rpm).

The clay samples, acquired by deep boring through the Boom clay formation at the Mol site, were first fragmented with a spatula, dried in a stove at 50 °C, crushed and sieved (100 mesh). Drying, crushing and sieving were repeated until nearly all clay passed the 100 mesh sieve, coarser fragments of sulfides, carbonates being removed partially by the sieving process. Finally the powder obtained was dried in a stove at 110 °C during 3 hours and stored afterwards in a closed teflon flask.

Radiation and thermal treatments were performed on samples prepared as described above.

Some samples refered to as "intact" samples, were only sieved without drying at 110 °C.

4.2. Results and discussion

Starting with the solutions prepared according to section 3.1 and after a stabilisation period of 45 days a number of batch experiments was carried out to investigate the adsorption of plutonium on clay as a function of depth of the layer. The time required to reach equilibrium was inferred from a series of equilibration periods ranging from 1 to 30 hours. The adsorption tends to stabilise after 16 hours. An equilibration time of 24 hours was then used for all subsequent batch experiments. The results are presented in figure 1.

The equilibrium data at concentrations in excess of 10 mg liter^{-1} and at a pH 1 to 3 show an increase in K_d with decreasing concentration which is consistent with an adsorption model by ion exchange near to saturation. The amount of plutonium adsorbed expressed in milliequivalents however exceeds by a factor of 4 to 5 the theoretical capacity of the clay which precludes adsorption as Pu^{4+} ions.

Very large K_d values are encountered when pH differences occur between initial an equilibrium solutions. They are certainly due to precipitation of the plutonium as insoluble hydroxides. Another group of data can be discerned with a pH in the range 7 to 9 showing a trend to increasing K_d values with decreasing equilibrium concentration. Apparently a pH between 7 and 8 is least favorable for adsorption to occur. For some data points this effect is masked by precipitation due to change in pH during equilibration.

Finally a group of points referring to the lowest concentrations are showing low K_d values because of the difficulties in accurately measuring the equilibrium concentrations at this low level.

The above considerations are merely a possible explanation for the data observed and no real understanding of the underlying phenomena and equilibria has yet been achieved. The fact that two K_d values exist for the same equilibrium concentration demonstrates clearly that other phenomena than ion exchange or adsorption are playing an important or even predominant role.

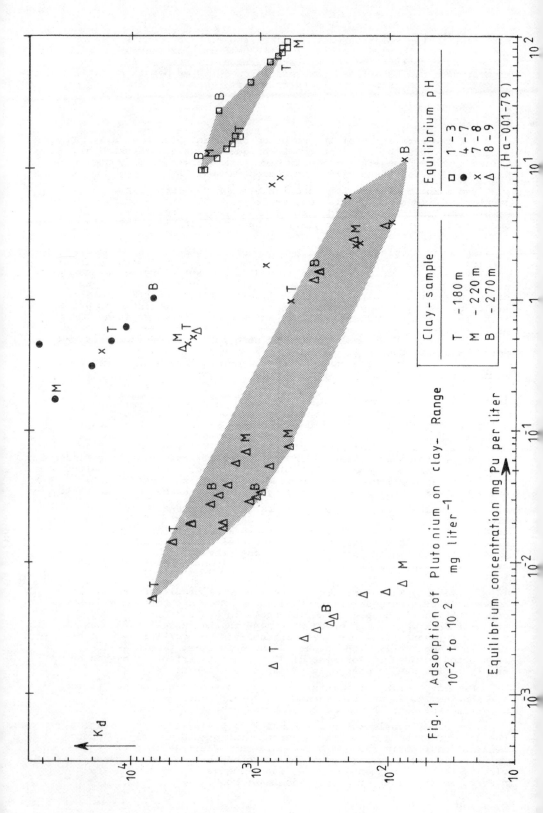

Fig. 1 Adsorption of Plutonium on clay- Range
10⁻² to 10² mg liter⁻¹

K_d

Equilibrium concentration mg Pu per liter

Clay-sample	
T	-180 m
M	-220 m
B	-270 m

Equilibrium pH	
□	1 - 3
●	4 - 7
x	7 - 8
△	8 - 9

(Ha-001-79)

Colloidal particles may add some scatter to the results but their effect should not be overestimated because of the large number of data showing similar trends. As far as the clay sampling layer is concerned there is a slight indication that the top and bottom layers are showing higher adsorption than the center of the layer although the effect is too small with respect to other possible causes to be clearly inferred from Figure I.

In order to obtain a better understanding and control more closely the different parameters involved a new set of experiments was carried out with starting solutions at pH 3. There are a number of data available indicating that the pH of the aqueous phase in the bulk of the clay layer is around 3.

However at the time those experiments where planned, it was not fully realised that the clay, after drying at 110 °C acquired neutralising properties to a variable degree. As a consequence high apparent K_d values were found corresponding to the values of Figure 1 where important changes in pH between initial and equilibrium solutions induced precipitation or coprecipitation of plutonium hydroxide.

The same experiments where also carried out with clay samples after a thermal treatment at 300 °C to investigate a possible degradation of the adsorption properties by heating due to the decay of high level radioactive waste. The effect, if any, however was completely masked by the precipitation of plutonium when the pH from the solution changed from 3 to a value between 6 and 7. Apparent K_d values were in the range 10^4 to 10^5 for equilibrium concentrations between 0.1 and 1 mg per liter and in the range 10^3 and 10^4 between 0.005 and 0.1 mg per liter. At the low concentration range the limited measurement sensitivity resulted in lower K_d values as already described with respect to Figure 1.

The lower concentration range has been investigated more in detail by using enriched ^{238}Pu as a tracer. The initial solutions covered a range from a concentration of 611 µg to 6.1 ng per liter. All were initially at a pH between 2 and 3 except one at a pH of 1.5. The results are given in Figure 2. The group at a pH between 1 and 2 results from the solution with initial pH at 1.5. Since the pH is essentially invariant the data may correspond to true ion exchange equilibria of an hitherto undefined soluble plutonium species. The distribution coefficients are relatively low because of the high protonic activity in the solution.

Another set of data where pH is nearly invariant is formed by the group within the pH-range 2 to 3. The K_d values are comprised between 700 and 3000 for samples taken from -140 m down to -250 m. The constancy of this K_d range over a four decade range in concentration is similar to an ion exchange model in the trace concentration region.

For the group of data with an equilibrium pH between 3 and 5, a moderate pH change has occurred and precipitation of plutonium hydroxide tends to yield higher apparent K_d values, in the concentration range 0.01 µg liter^{-1} to 10 µg liter^{-1}. At lower concentrations a decrease in K_d values is observed.

Equilibrium solutions with a pH between 5 and 7 still show higher apparent K_d values because of the very low solubility of plutonium species in this range. However the decrease in K_d values in the low concentration range sets on already downwards from 0.04 µg liter^{-1}. At an equilibrium pH higher than 7, a decrease in K_d values over the concentration range upwards from 0,01 µg liter^{-1} is observed whereas lower concentrations show a similar behaviour as the group of pH 5 to 7.

This behaviour is consistent with the existence of soluble anionic complexes which are not adsorbed on the clay and whose formation is pH dependent.

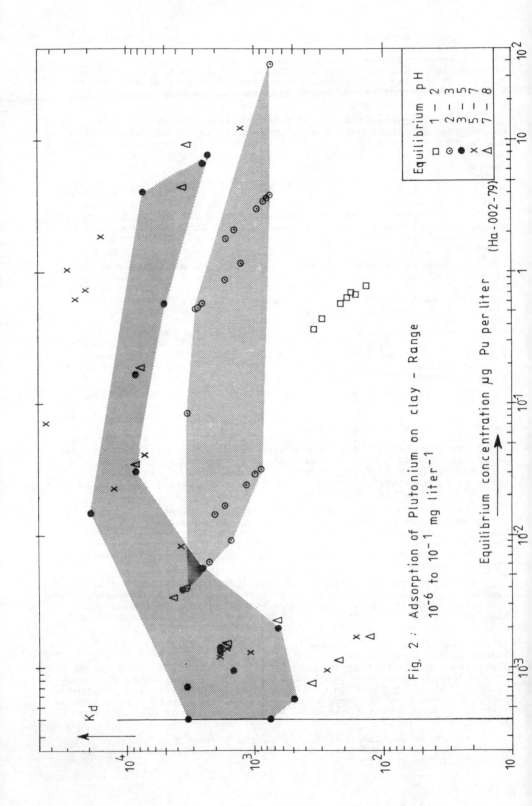

Fig. 2: Adsorption of Plutonium on clay – Range
10⁻⁶ to 10⁻¹ mg liter⁻¹   (Ha-002-79)

Equilibrium concentration μg Pu per liter

As the existence of $PuO_2CO_3OH^-$ and $PuO_2(CO_3)_2{}^{2-}$ has been reported [1] in the Eh and pH range of this work, and carbonate is the major anion in this ground water, it is tempting to explain the observations on this basis. A more detailed study of the plutonium species in solution however is necessary before such a conclusion can be drawn.

A limited number of experiments have been done with clay after thermal treatment up to 500 °C and after γ-irradiations up to a dose of 3.10^{10} Rad. The results are displayed in Table V. Again the degradation of ion exchange properties if any, due to radiation or thermal treatment is obscured by precipitation reactions. Only the clay heated at 500 °C and irradiated shows consistently small apparent distribution coefficients in spite of the pH changes. The reasons for this behaviour are not understood and further studies are required to investigate the importance of those effects with respect to plutonium migration.

5. SORPTION OF IODINE

Some investigations have been carried out concerning the behaviour of Iodine (as I^-) in solutions in contact with clay.

K_d-measurements were executed after mixing resp. Boom clay and resp. glauconite sand (both prepared as described in section 4.1 with $^{131}I^-$ traced solutions (Anversian water and dilute hydrochloric acid at pH = 3) of different iodide concentrations.

The results of the experiments are shown in Table VI and VII. An extremely weak sorption of I^- on both geological materials can be observed from de K_d-values (ranging from 5.12 to 0.18).

The K_d-values are independ of the I^- concentration and it is not excluded that other phenomena than chemisorption intervene.

6. CONCLUSIONS

The real distribution coefficients for plutonium in ground water can only be obtained in pH-Eh invariant systems. Otherwise precipitation reactions will result in very high apparent K_d values. A pH invariant system is difficult to obtain because of the neutralising effect of clay on acid solutions and the variation of this effect with depth of the clay layer.

Nevertheless K_d values can be expected to be at least 5.10^2 ml g^{-1} in a pH range from 2 to 9 in the absence of plutonium complexing agents. There is some evidence that at concentrations below 0.01 µg liter^{-1} a less adsorbable species exists whose formation is enhanced at higher pH.

It should be pointed out that in the case of an ion exchange mechanism the migration pattern of Pu follows the well established diffusion theory of soluble species in a porous medium. In the case of precipitations the mobility of the species is ruled by the laws of colloidal particles in a filtrating medium. The consequences of both migration patterns on this long term of geological disposal are very important and should be the subject of detailed investigation.

REFERENCE

[1] Brookins, D.G. : "Application of Eh-pH Diagrams to Problems of Retention and/or Migration of Fissiogenic Elements at Oklo" in "Natural Fission Reactors", (Proc. Technical Commitee Meeting, Paris 19-21, December 1977) p. 243-265, IAEA Vienna, 1978.

TABLE IV : Plutonium solubility in water containing
soluble components from the clay at -224 m

A : orginal water   B : boiled for 3 hours

	pH	Decanted solution	Centrifuged solution
A	2.22	16.8 $\mu g\ \ell^{1}$	20.1 $\mu g\ \ell^{-1}$
B	2.22	24.2	24.3
A	3.75	0.01	0.004
B	3.89	0.007	0.005
A	6.16	0.46	0.40
B	6.87	0.52	0.42

TABLE V : Adsorption experiments on clay after thermal and
irradiation treatment. Initial pH of Pu solutions was 3.

(CO = initial concentration $\mu g\ \ell^{-1}$; C_E = final concentration $\mu g\ \ell^{-1}$)

Sample		-220.6 m - -	-220.6 m - 3.10^{10} Rad	-220.6 m 24 h at 300 °C 3.10^{10} Rad	-220.6 m 24 h at 500 °C 3.10^{9} Rad	-240.6 m - - -	-240.6 m 24 h at 110 °C 3.10^{10} Rad
CO							
	pH	3.30	4.88	5.77	4.60	3.33	7.02
3247	C_E	47.4	39.5	3.73	1461	38.7	27.7
	K_d	3375	4065	40,500	61	4150	5819
	pH	4.36	6.92	7.28	6.48	3.86	7.72
299.5	C_E	2.08	0.86	< 0.01	134	2.35	6.59
	K_d	7170	17,342	> 10^5	61	6322	2223
	pH	4.76	7.07	6.27	7.03	4.60	7.63
139.9	C_E	3.87	9.01	< 0.01	38	1.55	5.66
	K_d	1757	726	> 10^5	138	4463	1185
	pH	5.03	7.48	7.47	7.07	5.15	7.66
46.2	C_E	1.15	1.14	< 0.01	20	6.11	3.92
	K_d	1930	1951	> 10^5	65	324	533
	pH	5.13	7.36	7.25	7.00	5.23	7.86
45.7	C_E	0.38	< 0.01	< 0.01	3.7	< 0.01	8.61
	K_d	6116	> 10^5	> 10^5	588	> 10^5	218
	pH	3.91	6.61	6.93	5.01	3.68	7.67
12.9	C_E	0.44	< 0.01	< 0.01	0.33	< 0.01	1.43
	K_d	1426	> 5.10^4	> 5.10^4	1893	> 5.10^4	402

TABLE VI : Sorption of I⁻ on clay samples

Initial solution I⁻ concentration (pH = 3, in mg/ℓ)		5	1	0.1
Sample				
1. glauconite sand (-140 m, intact)	K_d	1.63	0.70	1.85
	c_E	4.84	0.99	0.10
2. glauconite sand (-140 m, dried)	K_d	1.69	0.87	5.17
	c_E	4.84	0.98	0.09
3. Boom clay (-160 m, dried)	K_d	2.04	1.25	2.12
	c_E	4.8	0.98	0.096
4. Boom clay (-200 m, dried)	K_d	1.78	1.02	1.25
	c_E	4.8	0.98	0.098
5. Boom clay (-220 m, dried)	K_d	6.12	1.3	1.06
	c_E	4.8	0.98	0.1
6. Boom clay (-220 m, intact)	K_d	1.69	0.87	5.17
	c_E	4.84	0.98	0.09

TABLE VII : Sorption of I⁻ on clay samples high I⁻-concentrations

Initial solution concentration (in mg/ℓ, pH = 3)		1000	500	200	100	50	20	10
Sample								
Boom clay 200 m, dried	K_d	2.35	1	0.18	-0.34	1.76	1.19	1.21
	c_E	255	490	199	–	48.3	19.5	9.8

Discussion

D. RAI, United States

In your PuO$_2$ study, did you measure the pH after the initial adjustment at the beginning of the experiment ? In our work, similar to yours, the solubility of PuO$_2$ was very much a function of pH, which decreased continuously with time.

Do I understand rightly that you did not filter the samples ? We observed discrete particles of PuO$_2$ in our unfiltered samples. Do you think some of the variability in your equilibrium solutions may be due to this problem ?

P.P. de REGGE, Belgium

We did measure pH at the time the Kd experiments were started i.e. 45 days after preparation of the solutions listed in Table II. Remeasurement of the pH after several months did not show any significant change. Because of the large number of data showing similar trends the effect of solid particles, although obvious for some discrete data points, should not be overestimated. Solutions were indeed not filtered in the experiments performed until now. On the other hand, if plutonium is present as solid particles or in colloidal form its migration in that form should be considered as well, in addition to the adsorption studies and Kd determinations.

A. SAAS, France

Nos expériences menées à Cadarache montrent la présence de complexes organiques du plutonium. Ces composés organiques sont capables, selon les conditions du milieu, de produire des oxydations ou des réductions, ce qui modifie considérablement sa solubilité et sa dispersion dans le milieu. Des formes solubles complexes ont ainsi été mises en évidence dans des eaux superficielles naturelles et la dispersion "in situ" par les canaux d'irrigation confirme la présence de ces formes solubles à caractère physico-chimique variable.

P. BO, Denmark

Have you considered the instability of hydrogen-clays when doing experiments at low pH, in the range 1 to 3 ?

P.P. de REGGE, Belgium

In fact, the low pH range was observed in the clay itself. A series of samples through the layer results in pH values ranging from 3 to 7. At a pH lower than 2, indeed clays becomes instable but on the other hand it is not possible to obtain stable plutonium solutions with concentrations in excess of 0.1 mg/l without lowering the pH.

The equilibrium pH of the solutions was essentially between 2.5 and 8, and in this range the instability of clay is not an important parameter for the interpretation of the results.

D. RAI, United States

You mentioned that the observed solubility of PuO$_2$ is many orders magnitude higher than predicted on the basis of all Pu being present as Pu(V). We observed in our experiments on Pu(IV) hydroxide that the equilibrating solutions predominantly contained Pu(V). I wonder if presence of Pu(V) and discrete particles in the solutions can account for large variations in the observed solubilities of PuO_2.

P.P. de REGGE, Belgium

Until now we were only able to show the absence of Pu^{3+}, Pu^{4+} and PuO_2^{++} in the solutions but we are continuing our investigations to find out in what form plutonium is present. We believe that polymers of plutonium (IV) are still a possibility.

The presence of colloidal particles is certainly a possible explanation for some outlying data, but the large number of data showing similar trends seems to indicate that this effect should not be overestimated.

AN EXPERIMENTAL SET-UP FOR TRANSPORT MEASUREMENTS OF NUCLIDES
AT LOW WATER VELOCITIES IN CLAYS

A. van Dalen, J. Wijkstra
Netherlands Energy Research Foundation ECN
Petten (N-H), The Netherlands

ABSTRACT

Relative velocities of radionuclides in soils are generally calculated using distribution coefficients between groundwater and soil. In systems with low groundwater velocities the transport of nuclides showing slow distribution equilibria, this calculation method results in too high estimates of the relative velocity. The time to reach the distribution equilibrium of strontium in groundwater-clay systems is about 24 hours; the elution curve from a clay column shows considerable tailing at a groundwater velocity of 0.4 cm h^{-1}.

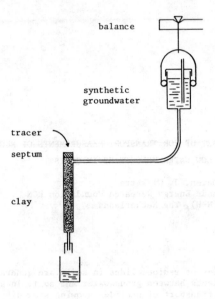

Figure 1. Schematic view of elution of tracers through a clay column.

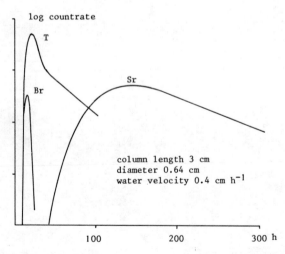

Figure 2. Elution of T, Br, and Sr with synthetic groundwater through a clay column.

INTRODUCTION

For the calculation of displacement patterns of nuclides in sub-soil experimental determined distribution coefficients are used together with data of the soil, e.g. density, porosity, and water movements. The measurement of a distribution coefficient is easily carried out. Distribution equilibria are not always established in short times, but by variations of contact time between the phases the equilibrium value can be determined. A drawback of the use of K_D values for the calculation of the displacement patterns is the fact that they are based on one value for each element under the specified conditions. The presence of more chemical states of the element in the system is not detected by measuring the distribution ratio. Fractions with deviating distribution ratios and thus other transport velocities are not remarked. This study was started to get more information about the behaviour of strontium because the distribution coefficients showed to be dependent on the contact times and ranged from 2 to 10 for equilibrium times of 1 to 24 hours. The same behaviour is expected for the element radium. The nuclear properties of Sr are, however, more favourable to carry out the experiments.

EXPERIMENTAL CONDITIONS

The Dutch subsoil in the northern part of the country consists mainly of deposits of clay material. Groundwater movements are very slow owing to a low hydraulic gradient and low permeability. Displacements of a few meters per year at a depth of 200-300 m are measured by C-14 dating methods [1].
At greater depths a salt layer is present with several saltdomes. The study is connected with the hypothetical event of the release of radionuclides from a repository of nuclear waste in a saltdome. The transporting groundwater in that case will be salt satured. The concentrations of the nuclides calculated from possible release rates will be very low compared with the exchange capacity of the clay material.

EXPERIMENTAL

A Dutch clay soil sample was obtained from the geologic department of the State University of Utrecht, without mineralogical specifications but sufficient homogeneous to carry out experiments with reproducible material. Samples of clay material from greater depths are available but only in small lots. The distribution coefficients of Sr are practically equal for both types of clay material. Synthetic groundwater was made using analytical data of water from wells at 300 m depth [2]. Sodium chloride was added to simulate the groundwater in the neighbourhood of a saltdome. The solution was 90 % satured to avoid problems of clogging by crystallization in the colomn. The amounts added per liter were 30 mg $CaCl_2$, 55 mg $MgSO_4.7H_2O$, 227 mg $NaHCO_3$ and 270 g NaCl. The pH of this solution was adjusted to 8.0±0.1 with HCl.
A clay column of 3 cm length and 0.65 cm diameter was used. The elution rate was measured by weighing the elutriant vessel (see Figure 1). For periods of a day and longer the elution rate proved to be constant within a few percent; larger deviations were measured during shorter intervals. The hydraulic constant (Darcy's law) was 5.3×10^{-4} cm min^{-1}.
Fractions were collected on time base. Two columns were prepared and the results with the ^{82}Br' tracer were reproducible: mean residence times of 5.0 h cm^{-1} and 5.2 h cm^{-1} were measured. ^{82}Br was produced by neutron irradiation of NH_4Br, tritiated water and ^{90}Sr-^{90}Y were obtained from The Radiochemical Centre, Amersham.
Gamma counting with a NaI(Tl) crystal was used for ^{82}Br determinations, liquid scintillation counting for T, and Cherenkov counting for ^{90}Sr after ^{90}Y has grown in.

RESULTS

The bromine tracer was chosen to serve as a tracer for the water displacement in the column. The shape of the elution curve showed practically no tailing. The mean residence time of the bromine tracer did not coïncide with the mean residence time of water calculated from the amount of water passing per unit of time and the cross section of the column using a porosity value of about 45 %.

Only the first arrival of activity from the column was in agreement with the
commonly used value for the porosity of clay.
The first activity of tritium arrived at about the same time as bromine; the
elution curve was only far from symmetrical, and showed considerable tailing
(see Figure 2). This behaviour was expected: around the clay particles is a layer
of strongly attached water molecules. Exchange of these water molecules with those
of the free flowing waterstream between the particles results in a retention of
the tritium tracer.
Much more pronounced retention was found in the elution of strontium. The K_D value
calculated from the first breakthrough of Sr and Br was 7 taking K_D of Br' zero.
This value is in the range measured by batch experiments. The retardation of Sr
relative to the water velocity is much larger than calculated according to the
K_D value. The last amount of ^{90}Sr activity left the 3 cm column six weeks after
injection.Only the elution front is moving according to the calculated velocity
but with a lower concentration. On the long range more and more strontium will
fall behind the front and the total effect is a lower velocity and a strong
spreading. More experiments have to be carried out before an attempt can be made
to give a more general and detailed picture of the most important processes determining the displacements of trace amounts: adsorption-desorption, kinetics of
equilibria, and diffusion in the liquid and solid phases.

DISCUSSION

The results of the column experiments give rise to the following remarks:
- a reliable tracer for the determination of the water velocity in clay materials
  is not available. Even in a short column with low water velocities bromine
  cannot be trusted. Tritium cannot be used as a tracer for water velocity due to
  exchange with water bounded to clay particles.
- the use of K_D values to calculate relative velocities of tracers showing slow
  exchange equilibrium, or high diffusion behaviour, leads to a too high estimate
  of the relative velocity. Also the tracer does not travel like a parcel but is
  considerably spread along the pathway.

LITERATURE

[1] Vogel, J.C.: Investigation of groundwater flow with radiocarbon. Isotopes in
    Hydrology, IAEA Symp. Vienna 1967, p.355-369.

[2] Meinardi, C.R.: Characteristic examples of the natural groundwater composition
    in The Netherlands. National Institute for Water Supply R.I.D. Note 76-1.

Discussion

G. LENZI, Italy

Avez-vous rencontré des difficultés dans les expériences de "leaching" en colonne d'argile ? L'argile provoque un phénomène de gonflement lorsqu'il est placé dans une colonne traversée par l'eau. Nous avons, nous aussi, travaillé avec des petites colonnes remplies d'argile montmorillonitique, et avons été confrontés à de terribles difficultés à la suite du phénomène de "swelling" de l'argile. D'autre part, le débit, à la sortie de la colonne est-il constant ?

A. van DALEN, The Netherlands

Swelling problems were not encountered, the clay material was only air dried before use. Filling of the column with synthetic ground water was performed in up flow at a very slow rate (1-2 weeks for a 3cm column). The flow rate during elution was measured by weighing the elution vessel. Drops leaving the column were protected with a small cap to avoid excess evaporation, thus no salt crusts were formed at the end of the tube.

L.R. DOLE, United States

J. Johnson and K. Kraus of Oak Ridge National Laboratory have been developing two experimental techniques to avoid the problem of maintaining long-term (days to weeks) column experiments with difficult clays. The problems of varing hydrodynamic column properties with time and ion exchange and the problem of precise chemical control for extended periods can be avoided by the two following experimental methods.

First, the addition of a porous filter aid with a low adsorption affinity for the migrating species will increase the flow through the column and result in shorter elution times.

Second, a spinning filter technique produces a radially fluidized bed, allowing rapid determination of adsorption and desorption rates. The results of these studies correlate well with static Kd results.

A. van DALEN, The Netherlands

Our aim was different : we tried to determine the relation between Kd values and transport in clay material for nuclides showing slow adsorption-desorption equilibria. Long duration of experiments is in this case inevitable.

P. BO, Denmark

Have you considered the effect of possible channeling on the Sr elution curve as compared to the Br elution curve.

A. van DALEN, The Netherlands

The formation of channels is always feared during the experiments, also because they were not visible. The only proof we have is the residence time of Br-82 in two different columns. Within the reproducibility of the experimental conditions the residence times were equal (5.0 and 5.2 h/cm) the same level of channeling looks improbable, so we concluded that this phenomenon was absent.

A.A. BONNE, Belgium

Est-ce qu'on a mesuré la teneur en strontium dans la carotte même ?

A. van DALEN, The Netherlands

The strontium content of the clay material has not yet been determined, it may be important for the exchange reaction of carrier free Sr-90 with the traces of strontium.

We noted a much faster elution of strontium when a small amount, but not carrier free, of reactor irradiated strontium was used (Sr-89).

A.A. BONNE, Belgium

La CEN/SCK (Mol, Belgique) a fait des essais semblables (strontium en milieu statique, solution à traceur sur la carotte d'argile, sans percolation) et on a constaté que la concentration en strontium dans l'argile même était beaucoup plus élevée, après l'expérience, que celle dans la solution à traceur. On en a conclu que, pour notre expérience, une précipitation de sel de strontium s'est produite.

C. MYTTENAERE, CEC

Cette remarque s'adresse aussi bien à cet exposé qu'à l'exposé précédent. Dans ces essais menés en laboratoire, durant des temps très longs et à des vitesses d'écoulement faibles, on peut se poser la question de savoir s'il est possible d'assurer que les conditions "biologiques" sont fidèles à celles présentes en grande profondeur (développement de micro-organismes durant l'expérimentation - modifications des formes chimiques initiales ...) ?

METHODE D'EVALUATION PREVISIONNELLE DES TRANSFERTS
D'UN POLLUANT DANS UNE NAPPE AQUIFERE
PREMIERS RESULTATS OBTENUS EN MILIEU SABLEUX
(CHANTIER DU BARP, GIRONDE)

Ch. Madoz-Escande, J-Ch. Peyrus
Commissariat à l'Energie Atomique
France

R E S U M E

Les études hydrogéologiques entreprises dans le cadre de
la sûreté radiologique des installations nucléaires ont notamment
pour but la prévision quantitative des conséquences de rejets accidentels de polluants radioactifs dans une couche aquifère (temps de
transfert, concentration à l'exutoire).
Cette prévision quantitative est obtenue à l'aide d'un
modèle mathématique à émission séquentielle dont l'emploi nécessite
la connaissance des paramètres physiques de la nappe et celle du
comportement du polluant vis-à-vis du milieu aquifère.
Les paramètres physiques d'un milieu poreux saturé sont
mis en évidence à l'aide d'essais par traceurs radioactifs sur maquette et aussi sur le terrain même.
Les premiers résultats obtenus sur un milieu sableux (chantier du BARP) sont présentés. La difficulté d'extrapoler au terrain
réel les conclusions des essais sur maquette a notamment nécessité
la réalisation d'un laboratoire mobile permettant des études in situ.
Le comportement du polluant vis-à-vis du milieu aquifère
est l'objet de recherches préalables en laboratoire sur les lois
d'adsorption intervenant entre le milieu et les radioéléments considérés sous différentes conditions de pH et de température.
Les résultats numériques obtenus demandent une confirmation sur le terrain ; une méthode en cours de développement sera
exposée, qui devrait permettre d'évaluer in situ les coefficients de
distribution.

METHOD OF FORECASTING POLLUTANT TRANSFER IN AN AQUIFER
INITIAL RESULTS OBTAINED IN A SANDY MEDIUM (BARP SITE, GIRONDE)

SUMMARY

The main aim of the hydrogeological studies undertaken in the context of the radiological safety of nuclear plants is the quantitative forecasting of the consequences of accidental releases of radioactive pollutants into an aquifer (transfer time, concentration at points of emergence).

This quantitative forecast is obtained with the aid of a mathematical model with sequential emission the use of which requires a knowledge of the physical parameters of the aquifer and a knowledge of the behaviour of the pollutant in relation to the water-bearing medium.

The physical parameters of a saturated porous medium are presented with the aid of radioactive tracer tests on a model and also in the field.

The initial results obtained in a sandy medium (BARP site) are presented. In view of the difficulty of extrapolating to field conditions the conclusions of tests on models, it was necessary to set up a mobile laboratory with which in situ studies could be undertaken.

The behaviour of the pollutant in relation to the water-bearing medium is the subject of preliminary laboratory research on the laws of adsorption operating between the medium and the radio-elements under consideration under different pH and temperature conditions.

The numerical results obtained call for confirmation in the field. A description is given of a method, now being developed, which should enable the distribution coefficients to be evaluated in situ.

1. INTRODUCTION

L'évaluation de la sûreté des différentes installations nucléaires (centrales, usines, centres de recherche, sites de stockage) nécessite notamment la mise en oeuvre de méthodes de calcul pour la prévision des transferts hydrogéologiques. Dans le cadre de son programme d'études, le Département de Sûreté Nucléaire du COMMISSARIAT A L'ENERGIE ATOMIQUE a entrepris de contribuer à la mise au point de tels modèles, en utilisant conjointement des méthodes de terrain et des méthodes de calculs, de manière à obtenir, pour des rejets donnés, des profils de concentration expérimentaux et théoriques simultanés dont la coïncidence est ajustée par un choix adéquat des coefficients de diffusion pour certaines valeurs données des paramètres./1/.

2. MODELE MATHEMATIQUE PREVISIONNEL DE TRANSFERT HYDROGEO-LOGIQUE

Pour calculer en fonction de l'instant d'arrivée l'évolution de la concentration d'un radioélément donné à un exutoire déterminé, on admettra ici dans un premier temps que l'écoulement à simuler par le modèle est horizontal, que l'aquifère est monocouche et homogène, que la vitesse de l'eau est constante et que, par conséquent, les coefficients de diffusion macroscopique sont proportionnels à la vitesse du fluide porteur. On tient compte aussi de l'adsorption des ions radioactifs par le terrain et de la décroissance radioactive du polluant constituant l'éventuelle contamination à traiter.

2.1. Equation générale de transfert

La concentration en polluant dans les phases liquide-solide d'un milieu poreux saturé et homogène varie avec le temps et la distance suivant la relation :

$$\frac{\delta C}{\delta t} = Kx \frac{\delta^2 C}{\delta x^2} + Ky \frac{\delta^2 C}{\delta y^2} + Kz \frac{\delta^2 C}{\delta z^2} - u \frac{\delta C}{\delta x} - \rho \frac{1-\emptyset}{\emptyset} \frac{\delta q}{\delta t} - \lambda \left(C + \rho \frac{1-\emptyset}{\emptyset} q \right) \quad (1)$$

où :

C : concentration volumique du polluant dans la phase liquide ($Ci.m^{-3}$)

Kx, Ky, Kz : coefficients de diffusion ($m^2.s^{-1}$)

q : concentration massique du polluant fixée sur la phase solide ($Ci.kg^{-1}$)

t : temps écoulé depuis l'émission du polluant (s)

u : vitesse de déplacement ($m.s^{-1}$)

x : distance depuis le point d'émission (m)

ρ : masse volumique de la phase solide ($kg.m^{-3}$)

$\emptyset$: porosité

λ : constante de décroissance radioactive (s^{-1})

L'équation (1) est l'équation primitive fondamentale de transport-diffusion.

Dans le second membre de cette équation, les trois premiers termes sont les termes "diffusion" ; le quatrième est le terme "convection"; le cinquième est le terme "rétention" et le dernier rend compte de la décroissance radioactive du polluant dans la phase liquide et sur la phase solide.

Si l'on admet que la loi d'échange du polluant depuis la phase liquide sur la phase solide est instantanée, réversible, alors $q = Kd.C$.

et /(1)/ devient :

$$\frac{\delta C}{\delta t} = Kx \frac{\delta^2 C}{\delta x^2} + Ky \frac{\delta^2 C}{\delta y^2} + Kz \frac{\delta^2 C}{\delta z^2} - u\frac{\delta C}{\delta x} - \rho \frac{1-\emptyset}{\emptyset} Kd \frac{\delta C}{\delta t} - \lambda C(1 + \rho \frac{1-\emptyset}{\emptyset} Kd)$$

d'où

$$\frac{\delta C}{\delta t}(1 + \rho \frac{1-\emptyset}{\emptyset} Kd) = Kx \frac{\delta^2 C}{\delta x^2} + Ky \frac{\delta^2 C}{\delta y^2} + Kz \frac{\delta^2 C}{\delta z^2} - u\frac{\delta C}{\delta x} - \lambda C(1 + \rho \frac{1-\emptyset}{\emptyset} Kd)$$

Si on pose

$$A = 1 + \rho \frac{1-\emptyset}{\emptyset} Kd \dots\dots\dots\dots$$ (terme retard constant
(dans le cas présent

on a :

$$\frac{\delta C}{\delta t} = \frac{Kx}{A} \frac{\delta^2 C}{\delta x^2} + \frac{Ky}{A} \frac{\delta^2 C}{\delta y^2} + \frac{Kz}{A} \frac{\delta^2 C}{\delta z^2} - \frac{u}{A} \frac{\delta C}{\delta x} - \lambda C \quad (2)$$

Si on pose en outre

$$\left. \begin{array}{l} Kx^* = \dfrac{Kx}{A} \\[2mm] Ky^* = \dfrac{Ky}{A} \\[2mm] Kz^* = \dfrac{Kz}{A} \end{array} \right\}$$
(pseudo coefficients de diffusion)

$$u^* = \frac{u}{A}$$
(pseudo vitesse du fluide porteur)

alors (2) devient :

$$\frac{\delta C}{\delta t} = Kx^* \frac{\delta^2 C}{\delta x^2} + Ky^* \frac{\delta^2 C}{\delta y^2} + Kz^* \frac{\delta^2 C}{\delta z^2} - u^* \frac{\delta C}{\delta x} - \lambda C \quad (3)$$

   Le modèle tridimensionnel de transfert hydrogéologique dont l'équation fondamentale a été présentée ci-dessus présente l'avantage de pouvoir tenir compte de la cinétique de l'émission et surtout de la géométrie de la source.

   Rappelons que ce n'est pas le cas pour un modèle monodimensionnel où seule la durée de l'émission est prise en compte.

   Quatre cas ont été envisagés :

- une émission ponctuelle instantanée,
- une émission ponctuelle non instantanée,
- une émission non ponctuelle instantanée,
- une émission non ponctuelle et non instantanée.

   2.2. Emission ponctuelle instantanée /2/

   Considérant que la concentration initiale dans le milieu est nulle, que le milieu est infini et que la concentration est nulle à l'infini, la solution de l'équation (3) pour l'émission ponctuelle instantanée d'une quantité W de radionucléide est :

$$C' = \frac{W}{8(\pi t)^{3/2}(Kx^* \, Ky^* \, Kz^*)^{1/2}} \, e^{-\left\{\frac{(x-u^*t)^2}{4Kx^*t} + \frac{y^2}{4Ky^*t} + \frac{z^2}{4Kz^*t} + \lambda t\right\}}$$

où C' désigne la concentration volumique du polluant dans le terrain : C' = CØA.

2.3. Emission ponctuelle non instantanée (fig.1)

A l'origine le polluant est injecté à la concentration Co, au débit Q, pendant l'intervalle de temps "d Θ" la quantité de polluant injectée est : Co Q d Θ
Pour une émission de durée Θ f, la solution de l'équation (3) est alors :

$$C' = \int_0^{\Theta f} \frac{CoQd\Theta}{8(\pi T)^{3/2}(Kx^* \, Ky^* \, Kz^*)^{1/2}} \, e^{-\left\{\frac{(x-u^*T)^2}{4Kx^*T} + \frac{y^2}{4Ky^*T} + \frac{z^2}{4Kz^*T} + \lambda T\right\}}$$

$T \neq t - \Theta$
Θ = date d'émission   ) rapportés à la même
    de chaque bulle   ) origine des temps
t = temps d'observation)

2.4. Emission non ponctuelle instantanée

Pour simplifier le raisonnement, on considère dans un premier temps une dimension de l'espace, par exemple l'axe Ox. On suppose que la source a une longueur "a" et qu'elle contient une quantité de polluant "W".
On a donc, C'o étant la concentration linéique du polluant dans le terrain (dans l'eau + sur la phase solide) :

      W = C'o.a
et   C'o = $\frac{W}{a}$

Maintenant, à cette source réelle, on associe une courbe de Gauss, ayant même maximum C'o et dont l'intégrale est la quantité W /2 /.

L'équation de cette courbe est la relation suivante :

$$C' = \frac{W}{(2\pi)^{1/2}\sqrt{2Kx^*_t}} \, e^{-\frac{(x-u^*t)^2}{4Kx^*t}}$$

Exprimant l'égalité des maximums il vient :

$a = (2\pi)^{1/2} . \sqrt{2.K^*x.t_{vx}}$

or $\sigma x = (2.K^*x.t)^{1/2}$, σx est une fonction de t au cours du transfert réel, on peut alors écrire :

$$\sigma x(t) = \left\{ 2K^*x \ (t + t_{vx}) \right\}^{1/2}$$

$$\text{ou} \quad \sigma x(t) = \left\{ 2K^*x \ (t + \frac{a^2}{4K^*x\pi}) \right\}^{1/2} \quad (4)$$

K^*x étant supposé indépendant du temps.

t_{vx} apparaît comme le temps fictif qu'il faut ajouter au temps réel de transfert depuis l'émission pour ramener le problème à celui d'une émission ponctuelle instantanée. Cela revient donc à considérer une émission ponctuelle instantanée "virtuelle" qui aurait été produite au temps "t_{vx}" avant l'instant d'émission "t = o" de la source réelle, mais qui donnerait, au temps de l'émission réelle, une distribution gaussienne d'intégrale W et de maximum Co.

Pour un problème tridimensionnel (source parallélépipédique de dimension a, b, c) le raisonnement est identique pour les autres dimensions. On aura donc :

$$\sigma y(t) = \left\{ 2K^*y \ (t + t_{vy}) \right\}^{1/2}$$

$$\text{soit} \quad \sigma y(t) = \left\{ 2 K^*y \ (t + \frac{b^2}{4 K^*y\pi}) \right\}^{1/2} \quad (5)$$

$$\text{et} \quad \sigma z(t) = \left\{ 2 K^*z \ (t + t_{vz}) \right\}^{1/2}$$

$$\text{soit} \quad \sigma z(t) = \left\{ 2 K^*z \ (t + \frac{c^2}{4 K^*z\pi}) \right\}^{1/2} \quad (6)$$

la solution générale devient alors :

$$C' = \frac{W}{(2\pi)^{3/2}\sigma x \sigma y \sigma z} \ e^{-\left\{ \frac{(x-u^*t)^2}{2\sigma x^2} + \frac{y^2}{2\sigma y^2} + \frac{z^2}{2\sigma z^2} + \lambda t \right\}} \quad (7)$$

C' désignant la concentration de polluant par unité de volume du terrain.

Un cas particulier intéressant pour nos essais est celui de la source cylindrique de rayon "R" et hauteur "C" (pollution émise à partir d'une portion de piézomètre) ; au lieu de

$$\frac{a^2}{4 K^*x \ \pi} \quad \text{et} \quad \frac{b^2}{4 K^*y \ \pi} \quad \text{on aura alors} \quad \frac{R^2}{4 K^*x} \quad \text{et} \quad \frac{R^2}{4 K^*y} \quad \text{dans les}$$

formules (4) et (5).

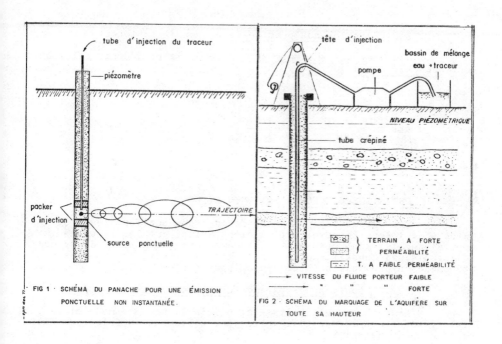

FIG 1 · SCHÉMA DU PANACHE POUR UNE ÉMISSION
PONCTUELLE NON INSTANTANÉE.

FIG 2 · SCHÉMA DU MARQUAGE DE L'AQUIFÈRE SUR
TOUTE SA HAUTEUR

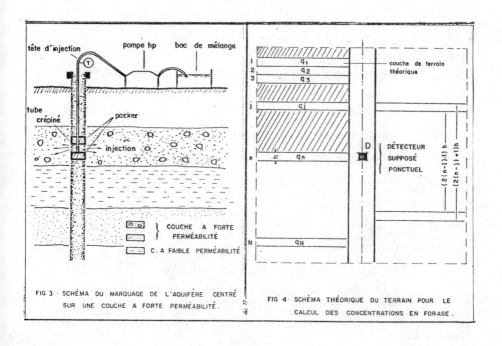

FIG 3 · SCHÉMA DU MARQUAGE DE L'AQUIFÈRE CENTRÉ
SUR UNE COUCHE A FORTE PERMÉABILITÉ.

FIG 4 · SCHÉMA THÉORIQUE DU TERRAIN POUR LE
CALCUL DES CONCENTRATIONS EN FORAGE.

2.5. Emission non ponctuelle et non instantanée

Cela revient à sommer sur le temps la précédente relation:
pour une émission de durée θf :

$$C' = \int_0^{\theta f} \frac{C_0 Q d\theta}{(2\pi)^{3/2} \sigma_x \sigma_y \sigma_z} \; e^{-\left\{ \frac{(x-u^*T)^2}{2\sigma_x^2} + \frac{y^2}{2\sigma_y^2} + \frac{z^2}{2\sigma_z^2} + \lambda T \right\}}$$

3. OBTENTION D'INFORMATIONS "IN SITU"

Les paramètres physiques de la nappe aquifère (sens et vitesse de l'écoulement essentiellement) ainsi que les réponses impulsionnelles sont obtenus simultanément par la méthode de traçage radioactif qui constitue un moyen de mesure directe in situ des phénomènes hydrodynamiques et hydrodispersifs.

3.1. Obtention des paramètres physiques de la nappe

La méthode multipuits est de loin la plus commode et la plus efficace malgré son prix de revient nettement supérieur à celui de la méthode unipuits.

Elle permet d'obtenir, entre deux points de l'espace, la fonction temporelle des concentrations, et de déterminer le sens d'écoulement local de la nappe aquifère.

D'une manière générale le mode opératoire est le suivant :

On injecte dans un puits une certaine quantité d'eau tracée qu'on laisse partir "au fil de l'eau" puis on mesure la concentration en traceur dans les puits d'observation.

Pour des raisons techniques l'expérimentation est réalisée en deux parties :
- la première consiste à "marquer" la nappe aquifère sur toute sa hauteur pour mettre en évidence les éventuels niveaux à circulation préférentielle (fig. 2),
- La seconde nécessite le marquage de la nappe aquifère par niveaux de mêmes caractéristiques hydrodynamiques. Cette technique est imposée par l'utilisation du modèle mathématique tridimensionnel où la connaissance de la géométrie de la source est indispensable (fig. 3).

Cette méthode présente des difficultés. L'expérience peut être longue si la vitesse du fluide porteur est faible et elle nécessite une étude préalable afin de déterminer la direction de l'écoulement pour diriger l'implantation du système piézométrique. Notons que si cette opération est plus onéreuse que la méthode unipuits, les résultats obtenus sont plus fiables et évitent un traitement mathématique complexe. /3/, /4/.

3.2. Obtention des concentrations des radioéléments

La nécessité de procéder au calage d'une réponse impulsionnelle théorique avec une réponse expérimentale pour obtenir les coefficients de diffusion macroscopique, implique la connaissance ponctuelle des concentrations "in situ" du radioélément utilisé comme traceur.

Pour des raisons évidentes de représentativité des résultats, la prise d'échantillons d'eau dans un forage en cours d'expérimentation n'est pas valable. Pour éviter ces inconvénients, nous nous sommes efforcés de mesurer les concentrations des radioéléments directement "in situ", mais comme les mesures effectuées sur le terrain se présentent sous la forme de pics photoélectriques de spectres gamma, elles nécessitent un traitement en vue d'obtenir les concentrations. Ce traitement est réalisé par un programme informatique spécial "FOCON" (Forage Concentration).

3.2.1. Principes de la méthode d'acquisition des concentrations à partir des mesures in situ

Le problème consiste à déterminer dans un forage le profil vertical de la contamination volumique d'un radioélément, dans un terrain constitué d'un ensemble de couches considérées chacune comme homogène, à partir de mesures in situ réalisées à l'aide d'une sonde (photomultiplicateur associé à un cristal INa) mobile à l'intérieur du forage.

Le rayonnement gamma émis par le radioélément contaminant est analysé par spectrométrie et seul le rayonnement non diffusé, c'est à dire correspondant au taux de comptage dans le pic photoélectrique du spectre, est recueilli et étudié en fonction de la profondeur /5/.

Pour un détecteur, un radioélément et un type de sol déterminés, le taux de comptage du détecteur est directement proportionnel au débit de fluence à l'intérieur du forage.

Ce débit de fluence résulte des différentes couches de terrain centrées sur le forage, chacune d'entre elles constituant une source à symétrie cylindrique d'axe vertical d'épaisseur déterminée et le libre parcours moyen du rayonnement gamma dans le sol est suffisamment faible pour que la concentration dans le cylindre source puisse être considérée comme homogène.

3.2.2. Formulation mathématique

Expression générale.
Le débit de fluence d'un rayonnement non diffusé au centre d'une source auto-absorbante et homogène et de symétrie cylindrique a pour expression (8) :

$$\psi_{(h)} = \frac{s}{\mu}\left[\int_0^\Omega \cos\theta \; e^{-a/\cos\theta} \; d\theta - \int_0^\Omega \cos\theta \; e^{-b/\sin\theta + c/\cos\theta} \; d\theta\right]$$

$\psi_{(h)}$ = débit de fluence pour une couche d'épaisseur h

s = nombre de photons émis par unité de volume

a = $\mu'r$

b = $\mu\dfrac{h}{2}$

c = $\mu r - a$

μ = coefficient d'absorption linéaire de la source

μ' = coefficient d'absorption linéaire du milieu situé au centre de la source

θ = angle formé par l'élément de volume et le plan horizontal

h = hauteur de la source

Ω = angle limite d'intégration déterminée par la hauteur de la source

r = rayon du cylindre central (Piézomètre)

En pratique, la source est constituée par une couche de terrain explorée par un détecteur mobile à l'intérieur d'un forage; il est donc nécessaire de tenir compte du tubage du forage, de l'enveloppe du détecteur et de la boue du forage, et l'expression (8) devient : fig. 4 /6/

$$\psi_{(h)} = \frac{s}{\mu_4}\left[\int_0^{\Omega}\cos\theta\; e^{-a/\cos\theta}\; d\theta - \int_0^{\Omega}\cos\theta\; e^{-b/\sin\theta + c/\cos\theta}\; d\theta\right]$$

(9)

avec :

$\psi_{(h)}$ = débit de fluence pour une couche d'épaisseur h

s = nombre de photons émis par unité de volume

Ω = arc tg $(h/_{2R})$

a = $\mu_1\; d_1 + \mu_2\; d_2 + \mu_3\; (R - \frac{D}{2} - d_2)$

b = $\mu_4\; h/2$

c = $\mu_4\; R - a$

μ_1 = coefficient d'absorption linéaire de la paroi (Fe) de la sonde

μ_2 = coefficient d'absorption linéaire du tubage

μ_3 = coefficient d'absorption linéaire de la boue ou de l'eau du forage

μ_4 = coefficient d'absorption linéaire de la couche contaminée (roche + eau)

R = rayon du forage

d_1 = épaisseur de la paroi de la sonde

d_2 = épaisseur du tubage

D = diamètre de la sonde

h = épaisseur de la couche considérée

L'application de cette formule couplée aux mesures " γ total" /7/ permet de connaître à différents moments de l'expérimentation les concentrations en radioélément dans le forage faisant l'objet des mesures.

3.3. Conclusions

Les expérimentations de terrain permettent d'obtenir les informations nécessaires au calage des modèles mathématiques de transferts hydrogéologiques.

Il est bien évident qu'en théorie ces expérimentations "in situ" semblent relativement simples à réaliser, mais en pratique il n'en est pas de même, car les terrains testés sont très souvent hétérogènes et donnent par conséquent des réponses délicates à traiter. C'est pour cette raison que dans la pratique nous "découpons" le terrain en strates de mêmes caractéristiques hydrodynamiques.

4. DETERMINATION DES COEFFICIENTS

Pour la prévision quantitative du transfert hydrogéologique, le modèle mathématique utilisé implique naturellement la connaissance des coefficients de diffusion et du comportement du polluant vis-à-vis du milieu aquifère.

4.1. Calcul des coefficients de diffusion

Il s'agit ici surtout de la diffusion macroscopique caracté
-risant le complexe hydrodynamique de l'aquifère considéré. Alors
que l'on possède une certaine expérience de la détermination des
coefficients de diffusion en laboratoire, leur détermination "in
situ" est un domaine relativement nouveau dont l'étude nécessite
la définition d'une méthodologie spécifique.

A partir d'une variation de concentration dans le temps en
un point donné de l'espace, correspondant à une pollution déterminée
on évalue les coefficients de diffusion macroscopique longitudinale
et latérale du milieu en utilisant des formules dérivées de la solution analytique de l'équation classique des transferts en milieu
fluide.
Les coefficients de diffusion Kx, Ky et Kz (Ky = Kz) sont calculés
par itération. Dans un premier temps, on recherche la valeur du
coefficient de diffusion transversale en se fondant sur la comparaison des intégrales sur le temps des concentrations en un point
(x, y, z) de la trajectoire du panache. Puis dans un second temps
on introduit comme donnée la valeur du coefficient de diffusion
transversale obtenue ci-dessus et on recherche la valeur du coefficient de diffusion longitudinale Kx par comparaison des formes des
réponses impulsionnelles expérimentales et théoriques.

4.2. Calcul des coefficients de distribution "in situ"

Lors de diverses expérimentations que nous avons réalisées
sur les sites, il s'est confirmé qu'il était difficile d'établir une
corrélation entre les coefficients de distribution (Kd) obtenu en
laboratoire et les résultats expérimentaux obtenus sur le terrain.
En ce qui concerne l'adsorption, les différences observées sont dues
essentiellement aux techniques de laboratoire /10/ où les milieux
à tester sont le plus souvent remaniés, leur structure ne présentant
plus qu'un lointain rapport avec le milieu originel.

Les corrélations étant délicates, sinon impossibles, nous
avons été amenés à entreprendre la mise au point d'une méthode de
calcul permettant d'évaluer le coefficient de distribution in situ
pour des conditions expérimentales simples. La méthode du double
marquage (injection d'un traceur idéal de l'eau (deutérium) et d'un
radioélément ayant des propriétés de fixation) permet d'établir soit
le rapport entre la vitesse de l'eau et la vitesse apparente de
l'ion, soit le rapport des coefficients de diffusion spécifiques
aux 2 réponses impulsionnelles obtenues au cours du double traçage.
Ce rapport donne directement le terme retard A de l'équation de
transfert.
Si cette méthode paraît intéressante en théorie, elle pose cependant de nombreux problèmes technologiques au niveau de l'appareil-
lage utilisé pour la détection du deutérium dans l'eau.

Il est évident que cette méthode n'offre pas la possibilité de résoudre les problèmes d'échange physico-chimiques entre
le terrain et le polluant mais elle intégrera tous les phénomènes
pour donner finalement un résultat global nécessaire aux calculs
prévisionnels.

5. PRESENTATION DES RESULTATS EXPERIMENTAUX

Pour disposer, dans un avenir proche, d'abaques permettant
d'obtenir directement les coefficients de diffusion macroscopiques
des expérimentations sont en cours sur une série de terrains "types"
dont la granulométrie est judicieusement choisie. Actuellement,
deux expérimentations ont été réalisées, l'une sur maquette métri-
que où le milieu est constitué par du sable fin tamisé provenant
de la sablière du BARP, l'autre sur le site de la sablière du BARP
(GIRONDE) où le squelette solide de l'aquifère est constitué par

un sable fin légèrement argileux /8/.

5.1. Présentation des résultats obtenus sur maquette

La maquette utilisée est constituée par du sable fin contenu dans une cuve cylindrique en acier inoxydable dont les dimensions sont d'une part 120 cm de diamètre et d'autre part 180 cm de haut. Le calage des différentes réponses impulsionnelles (fig. 5) expérimentales et théoriques, nous a permis d'obtenir les coefficients de diffusion intrinsèques longitudinaux et transversaux dont les valeurs sont respectivement $0,3$ cm $< \alpha x < 0,54$ cm et $1,9210^{-2}$cm $< \alpha y < 2,1910^{-2}$cm. La faible valeur de ces coefficients montre que le terrain testé est particulièrement homogène ; en effet dans le cas d'un milieu naturel non remanié, il est exceptionnel d'obtenir de telles valeurs.

5.2. Présentation des résultats obtenus sur le site du BARP

Le squelette aquifère du site du BARP est essentiellement constitué par du sable dont le diamètre moyen est de l'ordre de 0,3 mm. L'installation expérimentale est composée d'un rideau de pompage (5 puits) qui permet d'effectuer des expérimentations à vitesses variables ; d'un système piézométrique implanté perpendiculairement au rideau de pompage où s'effectuent les injections et les mesures. Le calage des réponses impulsionnelles expérimentales et théoriques a permis d'obtenir les résultats présentés dans les tableaux I et II.

TABLEAU I - Résultats de l'expérimentation du 24/10/78

Distance cm	U cm.s^{-1}	α_{cm}^{x}	α_{cm}^{y}
340	$7,80.10^{-3}$	2,31	0,65
640	$7,80.10^{-3}$	5,13	3,97
1000	$1,02.10^{-2}$	5,68	4,21

TABLEAU II - Résultats des expérimentations des 26/09/78, 24/10/78 et 14/11/78

U cm.s^{-1}	Distance cm	αx cm	αy cm
$2,6.10^{-3}$	340	2,69	2,44
$4,3.10^{-3}$	340	2,33	1,47
$7,8.10^{-3}$	340	2,30	0,65

Deux lois ont été mises en évidence :

La première montre une augmentation de la valeur des coefficients de diffusion intrinsèque longitudinale et transversale quand la distance de transfert augmente.

La deuxième montre que les valeurs des coefficients de diffusion intrinsèque longitudinale et transversale varient en sens inverse de la vitesse du fluide porteur.

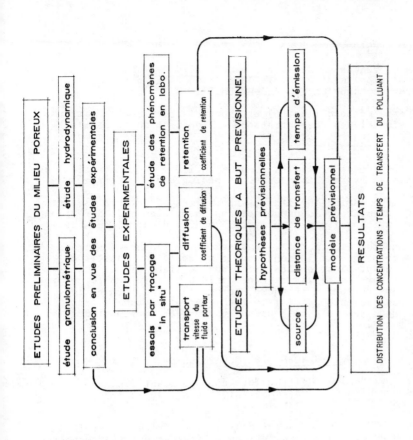

ETUDES PRELIMINAIRES DU MILIEU POREUX

étude granulométrique · étude hydrodynamique

conclusion en vue des études expérimentales

ETUDES EXPERIMENTALES

essais par traçage "in situ"

étude des phénomènes de retention en labo.

transport
vitesse du fluide porteur

diffusion
coefficient de diffusion

retention
coefficient de retention

ETUDES THEORIQUES A BUT PREVISIONNEL

hypothèses prévisionnelles

source · distance de transfert · temps d'émission

modèle prévisionnel

RESULTATS

DISTRIBUTION DES CONCENTRATIONS · TEMPS DE TRANSFERT DU POLLUANT

FIG 6 · organigramme descriptif d'une méthode d'évaluation prévisionnelle des transferts hydrogéologiques

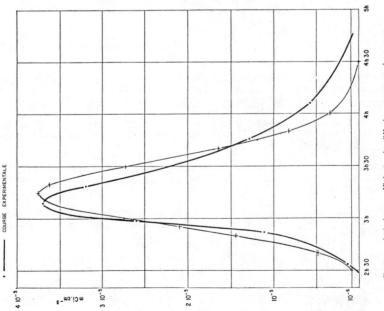

Manip maquette 10.2.77 13h 48   x = 110.0cm   W = 1,9 mCi

+ COURBE THEORIQUE
u = 9,9 10⁻³ cm.s⁻¹   Kₓ = 5,40 10⁻³ cm² s⁻¹   Ky = 2,15 10⁻⁴ cm² s⁻¹
σ x = 0,54 cm   σy = 2,17 10⁻² cm

· COURBE EXPERIMENTALE

fig 5 · calcul des coefficients de diffusion macroscopique par calage de la courbe théorique avec la courbe expérimentale

- 119 -

6. CONCLUSION

Nous venons de définir une méthode d'évaluation des transferts d'un polluant miscible dans l'eau d'une nappe aquifère et de montrer son application à deux cas réels.

Cette méthode (fig. 6) est fondée sur un modèle mathématique prévisionnel des concentrations en fonction du temps mis au point au Département de Sûreté Nucléaire et elle repose en outre sur un certain nombre d'opérations qui sont :

- Une étude préliminaire de l'aquifère (détermination des caractéristiques hydrodynamiques et morphologiques du terrain).

- L'obtention in situ des paramètres de terrain (sens et vitesse d'écoulement de la nappe) et des coefficients de diffusion et de distribution qui constituent les données d'entrée du modèle.

Dans son état actuel, le modèle prévisionnel des transferts hydrogéologiques tridimensionnels présenté ici constitue une première quantification des processus de transfert de pollution par les eaux souterraines. Cependant un certain nombre de problèmes délicats restent à résoudre, notamment une détermination fiable des coefficients de distribution (rétention sur la phase solide) qui nous semble devoir passer par un développement des études "in situ".

REFERENCES BIBLIOGRAPHIQUES

/1/ Escande Ch. :"Calcul du transfert d'un polluant dans une nappe aquifère - Programme TRIDISOL A1, A2 et 3", Rapport SESRS N°05 Janvier 1979.

/2/ Doury A. : "Méthode de calcul pratique et générale pour la prévision numérique des pollutions véhiculées par l'atmosphère", DSN - Rapport CEA R.4280. SACLAY 1976.

/3/ Peyrus J-Ch. : "Projet d'expérimentation hydrogéologique sur maquette", Rapport SESSN-R-11 - Septembre 1976.

/4/ Escande Ch., Peyrus J-Ch. : "Campagne hydrogéologique de sûreté dans le corail pour le site de MURUROA", Rapport SESSN R-08, Avril 1976.

/5/ Rhodes D.E., Stallwood R.A., Mott W.E. : "Intensity of unscattered gamma rays inside cylindrical self-absorbing sources", Nuclear Science and Engineering 9 - 1961 - pp. 41-46.

/6/ Hubbel J.H. : "Photon cross section, attenuation coefficient and energy desorption coefficients from 10 KeV to 100 GeV", NSRDS-NBS 29 - Août 1969.

/7/ Madoz J.P., Robert Th. : "Conversion des mesures in situ en concentration de polluant dans un aquifère", Rapport SESSN-R-05-Septembre 1976.

/8/ Peyrus J-Ch., Grimaud P. : "Etude hydrogéologique préliminaire du site du BARP (Gironde), Note Technique SESRS N° 26 - Octobre 1978.

/9/ Peyrus J-Ch., Geourjon M. : "Etude hydrogéologique de sûreté du site INFRATOME de LA HAGUE", Rapport SESR-R-30 - Septembre 1974.

/10/ Rançon D. : "Utilisation pratique du coefficient de distribution pour la mesure de la contamination radioactive des minéraux, des roches, du sol et des eaux souterraines", RapportCEA-R-4274-1972.

Discussion

J. GUIZERIX, France

Ma remarque a trait à la méthode du double marquage que vous nous présentez. Si vous disposez trois forages équidistants A, B, C sur une ligne de courant et si vous réalisez plusieurs injections que l'on puisse considérer brèves dans le premier forage A, avec des masses de traceur variant comme 1, 2, ... n, et si vous relevez en B les fonctions concentration-temps de la substance après interaction avec le milieu, vous avez gagné la moitié de votre pari si ces dernières sont représentées par des courbes affines de rapports 1, 2, ... n. Il faudrait encore vérifier que, si une injection brève est réalisée en A, la fonction concentration-temps observée en B, convoluée par elle-même donne à un facteur constant près la fonction obtenue en C. Je ne suis pas très optimiste sur les résultats parce que l'on a affaire à un système non linéaire : les fonctions concentration-temps obtenues en B ne permettent pas à mon avis de définir une réponse impulsionnelle.

P.Ch. LEVEQUE, France

J'ajoute quelques mots à la communication de Mme Madoz-Escande et de M. Peyrus.

La préparation de ces manipulations et des mesures en place a été basée, solidement, sur l'étude du milieu dans ses réalités naturelles. Ainsi, la transmissibilité a été rattachée à la hauteur réelle - ou épaisseur - de la couche perméable et à la perméabilité couche par couche.

Les coefficients de dispersion déterminés en laboratoire ont été comparés à ceux trouvés dans un milieu à symétrie axiale de 1,20 m de diamètre et 1,80 m de hauteur (cuve en acier inox).

Les caractéristiques des traceurs ont été longuement vérifiées in situ, avant la série des mesures.

Ainsi, la très bonne correspondance du modèle avec les mesures en place ne fait que sanctionner le souci constant de cette équipe, de ne s'éloigner en aucun cas, des paramètres naturels.

SURVEY OF RESEARCH PERFORMED IN THE EUROPEAN COMMUNITY ON THE MIGRATION
OF RADIONUCLIDES IN ROCKS (INDIRECT ACTION)

E. Della Loggia
Commission of the European Communities
Brussels

The papers presents a survey of the research on migration of radionuclides in the geosphere performed in the European Community Countries
in the frame of the Five year programme on Radioactive waste management
and storage of the Commission of the European Communities.

The investigation covers the migration of radionuclides in argilleous,
crystalline and salt rocks and its mathematical modelling.

The Nuclear Fuel Cycle and Power Industry Division for the indirect action of the Commission of the European Community and the Joint Research Centre of Ispra (direct action) are actively engaged in carrying out the Commission 5 years Programme on "Radioactive Waste Management and Storage".

As far as the indirect action is concerned the programme is carried out by contracts on an expenses sharing basis with public or private qualified industries or research organisation of the Community. Sixty two contracts have been signed to date for a total of 36.746.000 uc ($\simeq$ 46 millions of dollars) corresponding to a Commission participation of 16.585.000 uc ($\simeq$ 21 millions of dollars).

A consultative Committee of national experts nominated by their Governments assists the Commission for the management of the Programme.

Particular consideration is given to the problems connected with the migration of radionuclides in rocks which are candidate for use as a HLW repository and all the important aspects of the migration are subject of extensive research in the frame of contracts signed by the Commission with research organisation of the Community.

The geological formations under investigations include :

a) argilleous rocks (in Belgium, Italy, Denmark)

b) crystalline rocks (in France and in the United Kingdom)

c) salt rocks (in Germany and in Netherland).

whereas supplementary chemical studies are done at Joint Research Centre of Ispra and in Denmark.

Argilleous Rocks

The clay minerals of interest in connection with geological disposal of radioactive waste cover a wide spectrum of mineral structures and ion exchange properties.

The parameters which characterize the movement of ground water and the migration of radionuclides in this type of rock are investigated in Belgium where consideration is given to the possibility of building a repository for HLW in clay geological formations.

Due to the importance of ionic exchange for the retention of nuclides in argilleous rocks, it have been exclusively studied for the most important nuclides present in HLW.

Distribution coefficient (K_D) of fission products (Sr, Cs, Eu, and of plutonium have been measured in function of the pH, of the concentration of activity present in solution of the depth at which the rock samples were taken, of the ground water composition, of the temperature and of the dose of irradiation.

The experiments concerning Cs and Sr were performed using 12 samples of clay taken each 10 metres between -160 and -280 metre of depth and drive at 110°C. In order to approximate in the Laboratory the conditions of the natural environment, solutions of $CsNO_3$ and $Sr (UO_3)_2$ in ground water extracted at -150 m were used. The measurements were performed for several concentrations between 100 mgl^{-1} and 0.1 mgl^{-1}. Each experiment was made on a suspension of 250 mg of clay in 25 cc of ground water. The clay was then separated by centrifugation after the equilibrium had been (2 hours for Cs and 24 for Sr) and the activity of the solution determined.

Fig. 1 shows K_D of Cs and Sr   for samples taken at different depths for three different concentrations. The Ph of the solution was = 8. In order to see the influence of Ph, the same measurements have been repeated at different pH's and in Fig. 2 are shown the values of K_D obtained with solutions having a pH = 3.

The experiments have shown that in the case of Cs and Sr the sorption of the salts of Cs and Sr are little influenced by variation of Ph; in the case of Pu and Eu, on the contrary, the variations are significant.

The same technique has been used to study the properties of Pu and Eu salts : the experimental results however obtained for them are difficult to inter-pretate, but it is obvious that complexformation and precipitation reaction need to be considered nd possibly also mineraltransformations in some pH ranges. When the pH of the solution is < 3, the experimental K_D for Pu increases with the decrease of the Pu concentration in the solution : with Ph near 8, K_D decreases with the decrease of the Pu concentration. For Pu concentration < 1 gl^{-1}, the values of K_D fluctuates and it is impossible at the moment to find a satisfactory explanation for this. The results obtained for Eu follow the same pattern of Pu; there is a decrease of K_D with increasing concentration of the solutions. The static experiments are being supplemented with experiments performed using a dynamic technique which consists in filling a column of $\emptyset$ = 12 mm (for an height of 35 mm) with samples made by 3 gr. of clay dried at 110°C. The solution - at known concentration - of radionuclide with ground water is then passed through the clay column and the concentration measured at the outlet of the column.

With this method measurements have been performed for Eu and the first results seem to confirm the previous ones obtained with the static method. Other measurements are in progress for the same nuclide and for Pu.

Experiments have been performed to study the variations of K_D with the temperature for Cs, Sr and Eu and the influence of γ radiation on ionic exchange capacity of clay for Eu has been measured (Fig. 3). The effect of gamma irradiation seems similar to the effect caused by heating the samples, but the conclusions are not definite and further experiments are being performed.

Three contracts have just been signed by the Commission and Risø Research Laboratory covering important studies on the radionuclides migration phenomena.

Among the objects of the contracts is the investigation of the properties of ion exchange as a function of the concentration level and chemical composition of the acqueous phase and the evaluation of the influence of the particle size on ion exchange mechanism.

In order to study the permeability and solute dispersion and retention "in situ", single well technique will be used. The technique consists in injecting at a selected section in a well a pulse of a solution containing

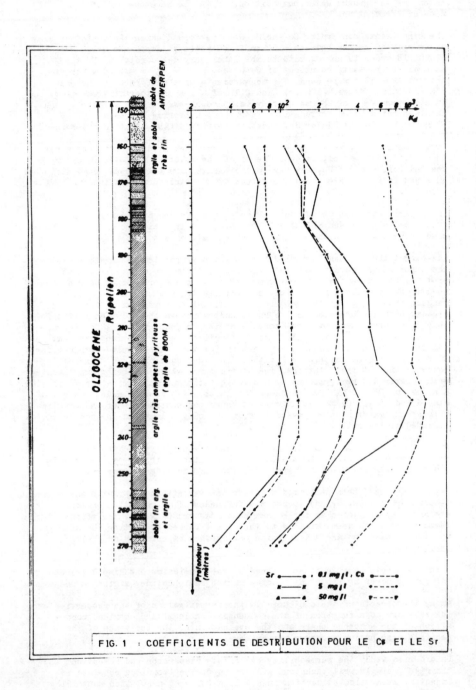

FIG. 1 : COEFFICIENTS DE DESTRIBUTION POUR LE Cs ET LE Sr

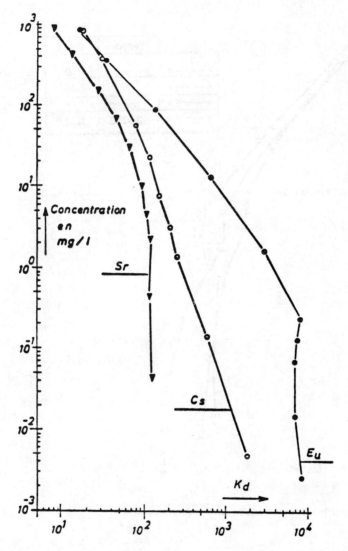

Fig. 2 : Variation of the distribution coefficient for variable concentrations of Sr, Cs, Eu (ph = 3)

(- 220 m clay samples)

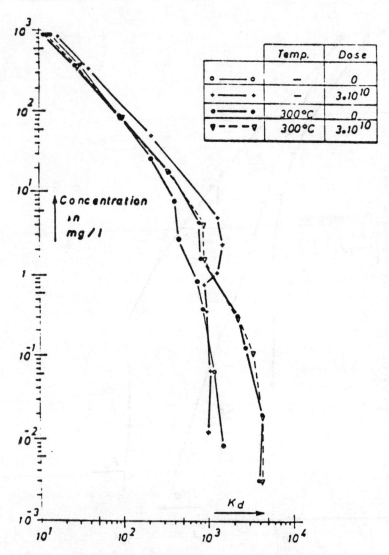

	Temp.	Dose
○ —— ○	—	0
• —·— •	—	3.10^{10}
● —— ●	300°C	0
▼ – – – ▼	300°C	3.10^{10}

Fig. 3 : Variation of the distribution coefficient for Eu
variable concentration
Influence of temperature and irradiation
(-220 m clay samples)

a) the isotope whose retention is to be measured (Cs, Sr, Eu) b) an isotope that will follow the movement of the water (as Br, Tc, Cr or negatively charged complexes) and washed deeper into the surrounding aquifer. Next the water is pumped back and the concentration profiles of the isotopes are measured by gamma spectrometry.

The "in situ" measurements will be supplemented by Laboratory determinations of ionic exchange capacity and distribution coefficients for Cs, Sr, and Eu on soil samples taken during drilling on the well.

Efforts are being devoted in Risø Laboratories to study the formation of carbonate and bicarbonate complexes which may strongly influence the migration phenomena by causing a substantial decrease in K_D of radionuclides with a corresponding enhancement of the migration of the species with ground water. The study concerns the characterisation of Carbonate and bicarbonate complexes with selected nuclides in the radioactive waste and an evaluation of the consequences for the migration of these nuclides through soil. The influence of carbonate complexing on migration phenomena will in addition be determined directly by chromatografic experiments.

The complexity of the processes involved in the transport of radionuclides and of the geological environment make the mathematical approach very attractive for studying the migration of the nuclides originating from leaching of active waste.

The departing point for the models developed in every country is the Fick's diffusion equation. The assumption and the basic parameters are characteristics for each site to which the model is applied.

In Belgium a three dimensional equation has been developed and it is being applied to study the effects and the relative importance of various physico-chemical processes affecting the migration in the porous media characteristic of the site.

The model, using as far as possible basic data obtained from the experiments, has been applied to the evaluation of the velocity of migration of actinides and fission products in the argilleous formation under consideration.

In particular the maximum concentration of Pu^{239} in drinking water as a consequence of Leaching and migration from the waste storage has been calculated.

Development is being carried out to the model in order to take into account geological (variation of K_D) and thermal inhomogeneity.

A computer programme "COLUMN" has been developed at Risø to solve a one-dimensional form of the transport equation.

The programme which solves the transport equation by finite difference explicite algorithms, can simulate Laboratory column experiments and describes the migration of radionuclides in saturated porous media.

The programme takes into account not only convection and dispersion in the Liquid phase but also precipitation and adsorption into the solid phase under non-equilibrium conditions. Ion exchange mechanism and radioactive decay are also included in the programme.

Crystalline Rocks

Granitic rocks are extensively studied in France and in the United Kingdom in view of their utilisation as waste HLW repository. The movement of radionuclides with ground water is being investigated in France by injecting into the rocks tracers (as I^-, Ce^-, No_3^-, Rodamine WT and uranine) and salts of Cs and Sr, through a borehole 40 m deep which contains a submerged pump capable of inducing convergent radial flow of the ground water. In the four cardinal directions satellite boreholes are drilled at approximately 12 m from the central point. The test consisted in constant rate pumping from the central borehole and short-durations injections into each of the satellite wells.

The transfer rates to the central well of the injected substances are interpreted by using a radial transfer model. With this method the hydrodinamic properties of the rocks have been determined, together with their porosity and with K_D for the various substances injected.

The measurements "in situ" have been checked with experiments carried out in Laboratory using both batch and columns methods on granite samples and they confirm the results obtained "in situ".

Particular attention has been paid in France to study the characteristics of natural and syntetic materials which can fix , due to their surface and structural properties, the radionuclides leached from a HLW repository and present in ground water. Their use is envisaged as barriers between the HLW repository and the granit rocks.

The minerals which theoretically present the most suitable characteristics for the fixation of radionuclides have been object of extensive research and their properties analyzed.

Twelve absorbent materials have been selected and their chemical composition particle size, exchange potential and mineralogical composition determined together with their absorption capacity. The variation of K_D has been studied in function of pH varying between 5 and 9. As a consequence of these measurements only 6 materials have been retained as candidate for artificial barriers and are object of further research. They are :

> Kaolinite
> Clinoptilolite
> Vermiculite
> Illite
> Attapulgite
> Bentonite

Experiments have been performed as well in order to study the effect of temperature variation on ionic exchange equilibrium for several nuclides (namely Cs, Sr, Np, Pu, Am). The use of tracers is also envisaged in the United Kingdom where considerable efforts are being devoted to investigate the use of uranium isotopes disequilibrium as a hydrological tracer and as a method of dating ground water. The dating methods are based upon the disequilibrium between the long lived nuclides and their shorter-lived daughters and they can indicate the residence time of formation water. Long residence time are usually associated with zones of low ground water velocity. Most natural waters are enriched in ^{234}U; the subsequent decay of the excess ^{234}U atoms should yield the age of the ground water system. More promising dating method is the one based on the evaluation of the values of ^{230}Th/^{234}U/^{238}U activity ratios in a given ground water system.

Efforts are being devoted to the development of this method in view of the fact that uranium isotope ratios and abundances can be used as natural tracers for determination of flow patterns and mixing proportions.

Factors affecting the dissolution and movement of radionuclides through the crystalline rocks are being investigated using Leachate from simulated Harwell glass containing plutonium and americium as well as specially prepared actinide containing solutions. Retention of the nuclides by powdered granite and by bentonite which is a likely candidate barrier material has been object of extensive research. The measurements, using samples of granite and strontium chloride solutions were carried out by column and batch experiments.

The experiments are being repeated at different temperatures and in presence of substances which may be contained in underground water sources as silica, chloride and fluoride.

The results of the measurements are reasonably satisfactory : K_D calculated from measurement of strontium taken up by granite show good agreement with one another. In the case however concerning the uptake of Pu from solution passed through grafite columns the values obtained are unreliable and work is going on to clarify the situation.

In the case of granitic rocks, the mathematical model used to describe the radionuclides migration is based as well on the Fick's law of diffusion.

The three-dimensional transport equation developed in France is composed by two modules : the first one is used to calculate the velocities and the hydraulic pressures and the second one to calculate the migration of the elements, taking into account the convection, the dispersion, the reversible absorption and radioactive decay.

Particular problem, when dealing with crystalline rocks, is the unhomogeneity due to the presence of fractures in the rocks. The representation in the transport equations of absorption in cracked media is introduced by using a coefficient of distribution through fracture, expressed as a quantity absorbed per unit of gross fracture area (ml/m^2) instead of the usual unit of mass of porous media (ml/g).

After having tested the validity of the model, the equations are at the moment being used for the interpretation of the tracers experiments described above.

In addition to the three dimension model a one-dimension model is used to evaluate the effects of the simultaneous displacement of harmful elements generated as a result of daughter-product formation during radioactive decay of the migrating elements (e.g. radium or uranium derived from plutonium, curium and americium).

The one-dimension transport equation used in the United Kingdom for migration studies is solved by use of the computer programme FACSIMILE . The equations are valid for a wide range of geologic materials from homogeneous particulate media, in which water flows through interstices, to faulted monolithic media, in which fissure flow is far greater than water movement through pores. Calculations have been performed to evaluate the discharge rate (curies/year), at the end of a 10 km rock column. Effects of varying Leach rate, dispersion coefficient, ground water velocity, sorption equilibrium constant and path length on the peak discharge of ^{99}Tc, ^{135}Cs, ^{226}Ra, ^{237}Np have been calculated.

Salt rocks

Migration of radionuclides from a HLW repository in salt formations, due to the characteristics of this type of rock is only thought of as a consequence of inrush of water in the waste repository, since the salt formations are free from circulating ground water. In the frame of the studies carried out in Germany to investigate the use of the Asse II salt Mine for storage of Radioactive Waste, experiments have been performed in order to ascertain the physical and chemical conditions of the originating solutions after a hypothetical water inrush. Some diffusion coefficients of radioactive nuclides have been measured within highly saturated brines.

The diffusion coefficients of $^{134}_{2}$Cs within saturated NaCl solutions are determined to be 7,54 ; 10^{-8} cm^2/sec, within brine sample from the Asse salt mine to be 9,7 $\dot{,}$ 10^{-5} cm^2/sec.

Experiments have been performed to measure the flow of water, but up to now no convection stream could be measured directly : the experiments are continued with more sensitive equipment.

A theoretical model is being developed for the representation of the effects consequent to a inrush of water in the HLW repository.

In the particular case of waste repository being studied in Netherland, the salt formations are surrounded by argilleous rocks. Due to this fact, K_D for samples of clay taken at various depths have been measured for Cr, Sr, Pu, Am especially to determine the influence of strong salt solutions which could be present in case of water inrush in the waste repository. A mathematical model has been developed to study the migration of radio-nuclides in this particular geological formation, in order to ascertain the consequences in the biosphere of a water inrush in the waste repository. The preliminary results obtained to date indicate that the expected release of nuclides will produce a contamination level lower than the present ICRP limits.

REFERENCES

1) Commission of the European Communities "The Communities R & D Programme Radioactive waste management and Storage". First annual progress report (1977) (EUR 5749)

2) Commission of the European Communities "The Communities R & D Programme Radioactive waste management and Storage" - Second Annual progress report (1978) (EUR 6128).

3) M.D. Hill and P.D. Grimwood "Preliminary Assessment of the Radiological Protection Aspects of Disposal of High-Level Waste in Geological Formations"
   NRPB - R 69   Harwell January 1978

4) J. Hamstra and B. Verkerk "Review of Dutch Geologic Waste Disposal Programme" Int. Conf. on Nuclear Power and its Fuel Cycle" - Salzburg - May 1977, paper IAEA - CN 36/289

Programme Sheet Nr 7 : Underground disposal

Con-trac-tor	Project and major programme items	Est.tot.expenditure (MUA)	EEC Contribution (MUA)	Time scale 1976 1977 1978	
GFK (D)	Disposal of radioactive waste in salt formations (The programme is going to be extended until 31.12.79)	2.81	1.34		015-76-1 WASD
ECN (NL)	Disposal of radioactive waste in salt formations (The programme is going to be extended until 31.12.79)	0.25	0.10		026-76-7 WASNL
CEN/ SCK (B)	Disposal of radioactive waste in deep clay formations (The programme is going to be extended until 31.12.79)	0.95	0.48		013-76-1 WASB
CNEN (I)	Disposal of radioactive waste in clay formations (The programme is going to be extended until 31.12.79)	1.40	0.70		020-77-9 WASI
CEA (F)	Disposal of radioactive waste in hard rock formations (The programme is going to be extended until 31.12.79)	1.60	0.70		019-76-7 WASF
UKAEA (GB)	Disposal of radioactive waste in hard rock formations (The programme is going to be extended until 31.12.79)	1.44	0.54		018-76-7 WASUK
AEC RISØ (DK)	Retention of radionuclides in geological formations (until 31.12.79)	0.87	0.42		038-78-1 WASDK
	Total	9.32	4.28		

Discussion

D.L. RANCON, France

Pour quelle raison avez-vous utilisé l'europium comme
élément de référence dans votre étude ?

V.E. della LOGGIA, CEC

Pour la raison suivante : dans l'éventail des produits
de fission, le CEN/SCK a sélectionné trois radionucléides typiques
pour chaque classe de valence : Cs (I), Sr (II) et Eu (III). Le
Eu (III) est donc considéré comme un élément représentant les tri-
valents.

D.L. RANCON, France

Dans les Figures 2 et 3, les Kd ont été mesurés pour des
concentrations pondérables supérieures à celles de radioéléments
purs sous des concentrations radioactives moyennes (mélange avec
des isotopes stables). Dans le cas du Cs-137 sous des concentrations
de l'ordre de 10^{-6} mg/l correspondant à des activités de l'ordre de
µCi/l, le Kd doit être beaucoup plus élevé.

V.E. della LOGGIA, CEC

Pendant les expériences, on a observé une croissance du
Kd inversement proportionnelle à la concentration de l'élément dans
la solution d'équilibre. Cette croissance permet en effet de prévoir
les valeurs de Kd plus élevées dans la région "traceurs purs".
Pourtant, dans la situation réelle, on n'aura pas seulement les
produits radioactifs dans les déchets mais également les isotopes
stables correspondants.

Au voisinage d'un repositoire, on peut prévoir ainsi,
dans le cas de débits d'eau extrêmement faibles comme nous en avons
dans l'argile, des concentrations nettement plus élevées que les
traceurs purs. C'est pour couvrir cette éventualité qu'on a déterminé les Kd dans une gamme de concentrations de 10^{-3} mg/l jusqu'à
10^3 mg/l.

Session III
Plutonium and Environment

Chairman — Président
Dr. D. RAI
(United States)

Séance III
Plutonium et environment

LABORATORY EXPERIMENTAL DEVELOPMENT FOR THE STUDY OF TRANSURANIC MIGRATION IN POROUS MEDIA

A. AVOGADRO, C.N. MURRAY and A. DE PLANO
Commission of the European Communities
Joint Research Centre - Ispra Establishment - Italy

ABSTRACT

A laboratory development for the study of transuranic migration in porous media is presented. The experimental system con-
sits of a water pathway which flows over the actinide bearing glass
and then through columns containing samples of typical deep soils.
The column outlets are continuously monitored and the soil contamination profiles are measured at the end of several months experiments. Parallel to these experiments an ultrafiltration system
and ion exchange resins are used in order to give an indication of
the size and charge of chemical species of actinides liberated
from the leached glass. Preliminary results and future developments
are reported.

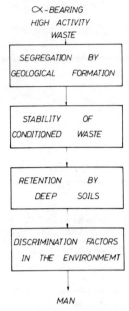

FIG.1   BARRIERS SYSTEM APPLIED
          TO HAZARD ANALYSIS

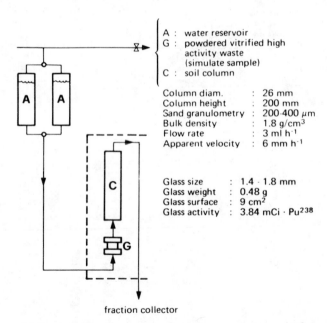

A : water reservoir
G : powdered vitrified high
     activity waste
     (simulate sample)
C : soil column

Column diam.	:	26 mm
Column height	:	200 mm
Sand granulometry	:	200-400 μm
Bulk density	:	1.8 g/cm^3
Flow rate	:	3 ml h^{-1}
Apparent velocity	:	6 mm h^{-1}

Glass size	:	1.4 - 1.8 mm
Glass weight	:	0.48 g
Glass surface	:	9 cm^2
Glass activity	:	3.84 mCi · Pu^{238}

fraction collector

Fig. 2 – Experimental Set-Up for Studying the Interaction of Glass
          Leachates with Deep Soil

In the framework of the assessment of the risk linked with the geological disposal of radioactive waste, an experimental programme studying the migration of transuranic nuclides in the geosphere is being developed by the J.R.C. Ispra.

The method adopted for the long-term risk assessment takes into consideration a set of natural and man-made obstacles which can prevent the migration of long-lived isotopes from the repository site to mankind. This "barrier system" consists of four barriers as illustrated in Figure 1. The behaviour of each barrier is being quantitatively assessed by probabilistic or deterministic methods.

The barrier which we are referring to today, the so called "geochemical barrier", is characterized by the retention capacity of the soil which is interposed between the geological formation selected for disposal and possible receiving aquifers.

We have thus supposed that as a consequence of an accidental event, water penetrates the repository and partially dissolves the actinide bearing glass, mouving afterwards through the surrounding porous formation.

In order to study the geochemical retention, we have reproduced in the laboratory as accurately as possible, the expected conditions of glass leaching and underground transport.

Borosilicate glasses containing Pu-238, Am-241 and inactive fission product oxides (Table 1) have been produced.

TABLE 1   GLASS MATRIX COMPOSITION

SiO_2	50	%
B_2O_3	15.67	%
Al_2O_3	5.16	%
Na_2O	12.50	%
waste oxides	16.67	%

The concentration of actinides simulates the composition of the vitrified high level waste after a cooling period of 1000 years. The standard production procedure of the glass consists of three successive fusions, the initial two fusions are carried out in nickel crucibles under a normal atmosphere, the last fusion is made in a graphite crucible in an inert atmosphere of argon at 1200°C. Radiochemical and metallographical examination of glass fragments showed homogeneous distribution of actinides and of metallic inclusions. The set-up adopted for the migration studies in soil columns is illustrated in Figure 2 ; a water pathway is established which flows over the actinide bearing glass and then through columns containing soil samples.

In Figure 2 is indicated an example of the method, and shows the experimental conditions adopted to study the elution behaviour of a leached solution passing through a pure sand column. Table 2 gives the composition of the water used in experiment which is typical of an aquifer overlying a clay formation. During the 35 days of the experiment, the outlet of the column was controlled. The results reported in Figure 3 indicate a continuous elution of Pu-238. At the end of the experiment, the column was cut into thin sections and the distribution profile of Pu-238 measured. The total quantity of Pu 238 found was 0.54 microcuries, corresponding to a glass leaching rate of 2.2×10^{-7} g cm^{-2} day^{-1}. It was found that 78% of the plutonium leached from the glass migrates apparently without interaction through the sand, while 22% is left in the top sections of the column.

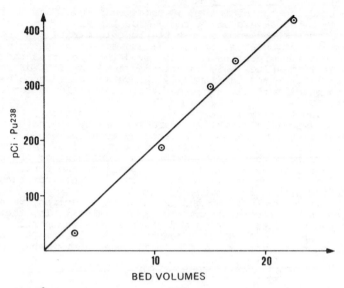

Fig.3ᵃ– Elution Behaviour of a ^{238}*Pu-Doped Glass Leachate on a Sand Column*

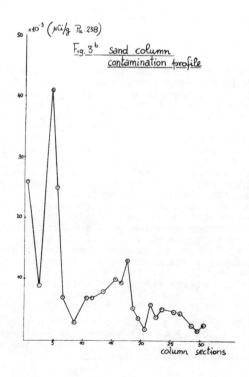

Fig.3ᵇ sand column contamination profile

TABLE 2   WATER COMPOSITION

Ca^{2+}	3.03	mg/l
Mg^{2+}	3.2	mg/l
Fe_{tot}	0.1	mg/l
Na^+	55.3	mg/l
K^+	7.71	mg/l
Cl^-	6.40	mg/l
SO_4^{2-}	0.5	mg/l
CO_3^{2-}	201	mg/l

pH = 8.35

Eh = 150   mV

In order to interpret the migration behaviour of actinides through soil columns, investigations of the chemical species of actinides produced by the leached glass have been started.

Figure 4 shows the set-up adopted for the initial experiments. A water pathway has been established which flowed over crushed and sized plutonium bearing glass and then through an ultrafiltration system followed by cationic-anionic resins.

During a 34 days experiment, the ultrafilter retained 20.1% of the leached activity while 60.6% and 19% were found on the cation and anion exchange resins respectively, 0.3% passed through the different components of the chain and was retained by the fraction collector.

In the case of americium bearing glass 89.2% of the activity was found on the filter while 10.7% and 0.1% were retained on the cation and anion exchange resins respectively.

A remarkably distinct behaviour was noticed between americium and plutonium leached from borosilicate glass both in the elution and filtration experiments. In order to systematically investigate the retention capacity of the geochemical barrier for the different actinides, a special set-up has now been installed in a glove-box. This installation (Figure 6) allows the carrying out of long term migration experiments with several columns filled with various typical sub-soils. At present characteristic porous formations overlying clay beds are under investigation.

Concerning the temperature influence on migration, the columns at present employed can be thermostated in the range 25 - 100°C. In order to study the influence of lithostatic pressure a simplified triaxial chamber for installation inside a glove-box has been built (Figure 5). Pressures up to 20 Kg cm^{-2} can be simulated.

This experimental programme aims to contribute to a better understanding of the mobility of actinides in deep soils, particularly in respect to their different physico-chemical forms.

Migration parameters through a soil column are normally calculated on the basis of distribution coefficients (K_d) which have been measured by equilibration experiments. This method, however, is difficult to apply in the case of a continuous input of a mixture of chemical species of the same isotope. Further it does

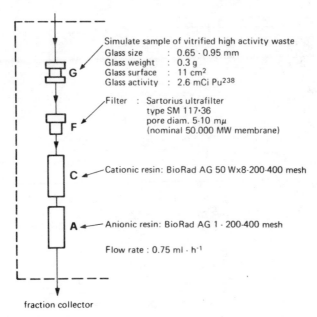

Simulate sample of vitrified high activity waste

G
Glass size	:	0.65 - 0.95 mm
Glass weight	:	0.3 g
Glass surface	:	11 cm^2
Glass activity	:	2.6 mCi Pu^{238}

F
Filter : Sartorius ultrafilter
type SM 117·36
pore diam. 5-10 mμ
(nominal 50.000 MW membrane)

C — Cationic resin: BioRad AG 50 Wx8-200-400 mesh

A — Anionic resin: BioRad AG 1 - 200-400 mesh

Flow rate : 0.75 ml · h^{-1}

fraction collector

Fig. 4 – Experimental Set-Up for Studying Chemical Forms of Glass Leachates

Fig. 5 Triaxial chamber for migration studies over high lithostatic pressure.

not take account thermodynamic and kinetic phenomena such as the
in-situ production of different chemical forms. The presence or
the possible continuous formation of more mobile forms, even at
very low concentrations, could result in a higher migration of the
isotope than would be predicted from a simple application of a
classical K_d. The theoretical development for the interpretation of
the experimentally observed results is reported by Dr. Saltelli in
two following papers.

The experimental methods that have been presented in this
paper have been developed for general geological porous media ;
for practical application of these techniques we would happy to be
put the JRC installation at the disposal of member countries in order that interested National Laboratories could use them in site-
-specific studies.

Fig. 6   Laboratory installation for migration experiments.

Discussion

L.R. DOLE, United States

Did you control the oxygen content of the elutant water ? Did you monitor the pH, Eh and ionic composition before and after contact with the glass and the columns ?

A. AVOGADRO, CEC

The input water is covered with a layer of an immiscible non polar liquid ; the oxygen dissolution in water is therefore avoided. The Eh and pH values are controlled before the column percolation.

The column is equilibrated during several days by percolating the water before the introduction of the actinide bearing glass in the water pathway.

G. LENZI, Italy

Avez-vous fait une interprétation géochimique, ou minéralogique sur les phénomènes de change-ionique dans le sable de votre colonne ? Et quelle était la composition minéralogique du sable ?

A. AVOGADRO, CEC

Dans cette expérience qui était seulement exploratoire, nous avons utilisé du sable de mer pur (Merck) ayant une capacité d'échange pratiquement nulle.

Seulement, les propriétés de filtration de la colonne apparaissent dans ce cas.

G. MATTHESS, Federal Republic of Germany

You have measured the granulometry of the sand, but did you measure the specific surface of the sand which is the parameter controlling the reactions between the solution and the solid phase ?

A. AVOGADRO, CEC

We have determined the specific surface of the glass, as reported in the paper (Fig. 2). Owing to the preliminary character of this experiment we have not measured the specific surface of the filtering column.

J. ROCHON, France

Avez-vous analysé tous les éléments dans les échantillons recueillis : concentrations des cations majeurs, anions, pCO_2, etc., paramètres qui régissent la distribution des espèces aqueuses ?

A. AVOGADRO, CEC

Nous avons contrôlé la composition chimique de l'eau d'alimentation, y compris le pH et Eh, uniquement avant la percolation à travers la colonne. A ce stade de l'expérience seulement la mesure des actinides est effectuée à la sortie de la colonne.

<u>V.E. della LOGGIA</u>, CEC

Have you got a definite plan for future experiments in your set-up and, if so, what is this plan ?

<u>A. AVOGADRO</u>, CEC

We have at the moment under study four columns containing the glauconite bearing sand coming from CEN, Mol, Belgium. We will continue to work with this material throughout 1979. For the future, as mentioned in my intervention, we would be happy to be involved in other site specific studies.

<u>A.A. BONNE</u>, Belgium

Est-ce qu'on a une idée de l'importance de la pression lithostatique pour la rétention ?

<u>A. AVOGADRO</u>, CEC

La pression lithostatique est prévue dans nos essais, dans le but d'éviter le gonflement de l'argile dans la colonne et pour maintenir l'argile sous forme de lentilles, dans les mêmes conditions de pression qui existent dans les formations profondes.

<u>A.A. BONNE</u>, Belgium

Vu la hauteur de la carotte de sable avec des lentilles d'argile, est-ce qu'on a une idée du temps nécessaire pour équilibrer la carotte avec la pression ?

<u>A. AVOGADRO</u>, CEC

Les lentilles d'argile sont dispersées dans le sable ; les dimensions de ces nodules sont très petites (< 0,5 cm). Nous pensons que l'équilibre de pression sera établi dans un temps très court.

"ASSESSMENT OF PLUTONIUM CHEMICAL FORMS IN GROUNDWATER"

A. Saltelli, A. Avogadro, G. Bertozzi
Commission of the European Communities – J.R.C. Ispra (Italy)

ABSTRACT

The leaching of plutonium from vitrified high-level waste can result in very low plutonium concentrations, of the order of 10^{-8}M, in the groundwater, where plutonium is shared among different complexes of the ions Pu^{4+}, Pu^{3+}, PuO_2^+, PuO_2^{++} with the anionic components of the medium.

Thermodynamic information is available in literature on some equilibrium constants, so that it is possible to describe quantitatively the plutonium distribution among different chemical forms, for a given water composition.

Pu (IV) complexes with carbonate ions are known to be very stable; however, the numerical value of 10^{47} which has been attributed to the stability constant appears to be too large by many orders of magnitude; a parametric study of that constant has been performed.

1. INTRODUCTION

The plutonium content of vitrified high-level waste remains at important levels for several thousands of years after the decay of the fission products; plutonium becomes therefore one of the elements which govern the risk linked to a possible leaching of the wastes by flowing groundwater [1].

A good degree of knowledge of the physico-chemical properties of leached out plutonium in groundwater is a major premise in order to allow a reliable assessment of the migration processes which could bring plutonium towards the biosphere because groundwater is the most likely vehicle for this transport.

The purpose of this paper is to provide a preliminary definition of the plutonium distribution among its different oxidation states, chemical species, and complexes, on the basis of the thermodynamic information at present available; further, the most important parameters which require better understanding are underlined.

Plutonium in natural waters may exist in different oxidation states [2,3,4,11], exhibiting radical differences in chemical behaviour; valence state (IV), the most stable under a large range of environmental conditions, is extremely insoluble and exhibits a very strong tendency towards polymer formation; valence states (III), (V) and (VI) are generally more soluble and do not polymerize; valence state (V), which previously was considered unstable, may be in reality the species which governs the total plutonium solubility, at low plutonium concentrations and in absence of complexing agents. In complexing environments species (III), (IV) and (VI) compete in determining the solubility, as a function of pH, redox potential and complexing features of the solution.

As a matter of fact, the general problem of plutonium speciation in natural waters may be as yet considered unresolved. Several efforts have been made to determine the plutonium speciation in sea water and fresh water; in groundwater, however, the problem is simplified due to the absence of organic matter (organic complexing agents, micro-organisms, etc.).

Attempts have been made by Polzer [4], Silver [5,6,7,8] and others to treat thermodynamically the general problem of plutonium speciation. An effort was also made to draw diagrams for various plutonium valence states, as a function of pH and Eh using the available thermodynamic data $\lfloor 3,9 \rfloor$. The lack of data for some complexes, and the uncertainties of stability constants for others, make it difficult to extend such calculations to environmental situations. However, it is worth noting that, among the many complexes which are possible in a given situation, only a few are of importance in determining the plutonium chemical species.

Rai and Serne $\lfloor 9 \rfloor$ have studied theoretically the stability of plutonium complexes in natural waters containing carbonate ions, and they reported that Pu (III) and Pu (VI) are the most important valence states in such solutions. These authors showed that, at low Eh values, Pu (III) carbonates are the predominant species, while Pu (VI) complexes become the most important components at high Eh values. However, it is to be stressed that Rai and Serne neglected Pu (IV) complexes.

Polzer $\lfloor 10 \rfloor$ performed an experimental study on plutonium sorption on different soil types : he identified that PuO_2^+ ion and, to a minor degree, Pu (VI) carbonates are the species responsible for the overall plutonium solubility; this author also considered the formation of Pu (IV) carbonates to be of minor importance. In our opinion, this assumption may be misleading because these complexes seem to be very stable.

Andelman and Rozzell $\lfloor 11 \rfloor$ published solubility curves for various plutonium oxidation states, taking into consideration both solutions with and without carbonates. These authors used for the Pu (IV)-carbonate stability constant the value reported by Moskviz and Gel'man $\lfloor 12 \rfloor$ of 10^{47} for a complex of the type $\lfloor Pu (CO_3) \rfloor^{++}$.

In some works published from 1970 to 1972, Silver $\lfloor 5,6,7,8 \rfloor$ developed a method for assessing plutonium speciation; it consists of describing the total complexing behaviour of the solution using an "alpha-factor", defined as the ratio of the concentration of all the soluble forms for a given valence to the concentration of the free uncomplexed ion.

2. PLUTONIUM SOURCE TO GROUNDWATER

It is current opinion that the leaching of plutonium from PuO_2 pellets results in the formation of a polymeric plutonium hydroxide phase [2,13,14]. However, plutonium concentration in vitrified high-level waste is not large; on the contrary the leaching rate of the glass is very low, so that the expected plutonium concentration in flowing groundwater may be evaluated to fall in the range of $10^{-8} - 10^{-9}$ M, as a function of several parameters, such as the specific surface of the glass and its overall quantity contacted by groundwater, groundwater velocity and temperature, etc. At such low concentrations, the probability of polymerization reactions is very small; therefore, in this case we have assumed that plutonium is present in solution only in the form of soluble complexes and free ions. Of course, colloidal silica migrating in groundwater can bear some of its original plutonium content; this phenomenon is not considered in the present study.

3. AVAILABILITY OF THERMODYNAMICAL DATA

In order to define the nature of soluble plutonium forms, we used Silver's method to calculate alpha factors from available thermodynamical data. When data were not available, the best possible estimation was made.

Information available in literature [15, 19] on complex formation with different complexing agents is summarized herewith; the stability constants are reported in Table 1.

Two complexes of Pu (III) with Cl^- ions are reported in the literature : [Pu Cl]$^{++}$ and [PuCl$_2$]$^+$; Pu (IV) also forms two different complexes : [PuCl]$^{3+}$ and [PuCl$_2$]$^{++}$. For Pu (V), the uncharged complex [PuO$_2$Cl]^0 is reported, while for Pu (VI) [PuO$_2$Cl]$^+$ and [PuO$_2$Cl$_2$]^0 are known. The stability constants of all these complexes are rather low, so that only at very high Cl^- concentrations do they need to be taken into consideration.

More stable complexes do exist between plutonium (IV) and (VI) and F^- ions; however, fluoride ions are very scarce in natural groundwaters, so that they can easily be neglected.

Sulphate ions form rather stable complexes with Pu (IV), of the type [PuSO$_4$]$^{++}$ and [Pu(SO$_4$)$_2$]^0; they could be of some relevance in sulphate-rich waters. Sulphate complexes of Pu (III) seem to be of minor importance. Data on sulphate complexes of Pu (V) and (VI) are not available, even though there is some qualitative evidence of complex formation for Pu (VI).

Complexes with carbonate ions are considered to be the most stable among inorganic complexes of plutonium, so that when carbonates are present, complex formation with other anions often becomes negligible. Unfortunately, substantial data are not yet available or are not reliable; a good knowledge of such data would be of crucial importance in assessing the fate of plutonium in natural waters, where carbonates are always present. For Pu (IV), the following equilibrium was assumed by Moskvin and Gel'man to exist in concentrated carbonate solutions $\underline{/}12\underline{/}$:

$$Pu^{4+} + CO_3^{--} \rightleftharpoons \underline{/}PuCO_3\underline{/}^{++}$$

The value of the corresponding constant $K_{(IV)}$ has been calculated by these authors, who quoted a value of $= 10^{47}$; this is the only experimental value available in literature and recently it has been strongly questioned $\underline{/}2, 4, 11, 16\underline{/}$.
On the basis of theoretical considerations, Polzer $\underline{/}4\underline{/}$ has suggested a value ten orders of magnitude lower. In reality, it is easy to see that a constant of 10^{47} would lead to unacceptable conclusions on the solubility of plutonium; for instance, in a natural system in equilibrium with atmospheric CO_2, the carbonate concentration would be sufficient to avoid any precipitation of Pu $(OH)_4$; in fact, huge concentrations of plutonium carbonate would be necessary to generate a Pu^{4+} concentration sufficient to reach the solubility product of the plutonium hydroxide (10^{-56}). On the contrary, experimental studies on plutonium hydroxide solubility in aqueous systems in contact with atmospheric CO_2 have shown values which, while exceeding the solubility product of 10^{-56}, remain at much lower levels in comparison with those expected on the basis of the carbonate stability constant of 10^{47} $\underline{/}17\underline{/}$. Probably, Moskviz and Gel'man's measurements also included as soluble plutonium a part of low molecular weight polymer, stabilized by the high carbonate concentration employed in the experiments. The same authors suggest that the complex could be of the type : $\underline{/}Pu(OH)_nCO_3\underline{/}^{2-n}$; Casadio and Orlandini in a recent polarographic work indicate for n a value of 4, at pH = 10.5 $\underline{/}18\underline{/}$.

For Pu (VI) several complexes with carbonates are reported in literature $\underline{/}15, 19\underline{/}$:

TABLE 1 : Equilibrium constants for some plutonium complexes.

Reaction	log K	
$Pu^{3+} + Cl^- \rightleftharpoons PuCl^{2+}$	-2.4	
$Pu^{3+} + 2Cl^- \rightleftharpoons PuCl_2^+$	-5	
$PuO_2^+ + Cl^- \rightleftharpoons PuO_2Cl^O$	-0.17	
$Pu^{4+} + Cl^- \rightleftharpoons PuCl^{3+}$	0.14	
$Pu^{4+} + 2Cl^- \rightleftharpoons PuCl_2^{2+}$	-0.17	
$PuO_2^{2+} + Cl^- \rightleftharpoons PuO_2Cl^+$	0.09	
$PuO_2^{2+} + 2Cl^- \rightleftharpoons PuO_2Cl_2^O$	-0.4	
$Pu^{4+} + F^- \rightleftharpoons PuF^{3+}$	6.77	
$Pu^{3+} + SO_4^{2-} \rightleftharpoons PuSO_4^+$	1.26	
$Pu^{3+} + 2SO_4^{2-} \rightleftharpoons PuSO_4^-$	1.70	
$Pu^{4+} + SO_4^{2-} \rightleftharpoons PuSO_4^{2+}$	3.19	
$Pu^{4+} + 2SO_4^{2-} \rightleftharpoons Pu(SO_4)_2^O$	4.71	
$Pu^{4+} + CO_3^{2-} \rightleftharpoons PuCO_3^{2+}$		47
$PuO_2^{2+} + CO_3^{2-} \rightleftharpoons PuO_2CO_3^O$		12
$PuO_2^{2+} + 2CO_3^{2-} \rightleftharpoons PuO_2(CO_3)_2^{2-}$		15
$PuO_2^{2+} + 3CO_3^{2-} \rightleftharpoons PuO_2(CO_3)_3^{4-}$		18.3
$PuO_2^{2+} + OH^- + CO_3^{2-} \rightleftharpoons PuO_2.OH.CO_3^-$		23.8
$PuO_2^{2+} + 2OH^- + CO_3^{2-} \rightleftharpoons PuO_2(OH)_2CO_3^{2-}$		23
$PuO_2^{2+} + HCO_3^- + CO_3^{2-} \rightleftharpoons PuO_2.HCO_3.CO_3^-$		12
$Pu^{3+} + H_2O \rightleftharpoons PuOH^{2+} + H^+$	-7.22	
$Pu^{4+} + H_2O \rightleftharpoons PuOH^{3+} + H^+$	-1.51	
$PuO_2^{2+} + H_2O \rightleftharpoons PuO_2OH^+ + H^+$	-3.33	

$\angle\bar{}PuO_2(CO_3)_3\bar{\mathcal{J}}^{4-}$, $\angle\bar{}PuO_2(CO_3)_2\bar{\mathcal{J}}^{--}$, $\angle\bar{}PuO_2.OH.CO_3\bar{\mathcal{J}}^-$,

$\angle\bar{}PuO_2(OH)_2CO_3\bar{\mathcal{J}}^{2-}$; $\angle\bar{}PuO_2.HCO_3.(CO_3)\bar{\mathcal{J}}^-$.

In the groundwater chemical conditions, $\angle\bar{}PuO_2.OH.CO_3\bar{\mathcal{J}}^-$ appears to be the most stable complex (see table 1).

No quantitative data are available on complexes between Pu (III) and carbonates, but qualitative evidence is reported $\angle\bar{}19\bar{\mathcal{J}}$. Plutonium (V) complexes do not seem to play an important role.

Soluble complexes with hydroxyl ions are known. For Pu (III), the form $\angle\bar{}PuOH\bar{\mathcal{J}}^{++}$ is reported, while Pu (IV) and Pu (VI) give $\angle\bar{}PuOH\bar{\mathcal{J}}^{3+}$ and $\angle\bar{}PuO_2.OH\bar{\mathcal{J}}^+$, respectively ; Pu (V) exhibits little tendency towards hydrolysis.

4. ESTIMATION OF UNKNOWN CONSTANTS

As already said in the previous paragraph, the carbonates are the most important inorganic agents in governing the plutonium solubility in natural water. However, there are some difficulties in estimating alpha factors for the various plutonium oxidation states :

I) — lack of data on Pu (III)— carbonate complexes,
II) — unreliability of Pu (IV)—carbonate stability constant.

In order to overcome point I), the following estimation has been made : it is commonly accepted that the complexing power of the anions of dibasic acids towards plutonium follows the order :

$$\text{a)} \quad CO_3^{--} > C_2O_4^{--} > SO_4^{--} \quad ;$$

on the other hand, Pu ions complexability follows the order :

$$\text{b)} \quad Pu^{4+} > Pu^{3+} \geqslant PuO_2^{++} > PuO_2^+$$

The values of the constants for oxalate complexes of Pu (III) are known $\angle\bar{}19\bar{\mathcal{J}}$; thus we have assumed that stability constants of carbonate complexes are not lower than those of oxalates. The stability constants for dioxalate and dicarbonate complexes of Pu (III) and Pu (VI) are given in Table 2. It may be seen that carbonate and oxalate complexes for Pu (VI) follow series a); on the other hand, Pu (III) oxalate is less stable than Pu (VI) oxalate : this inversion in series b) may be explained on the

TABLE 2 : Stability constants for dioxalate and dicarbonate complexes of Pu (III) and Pu (VI).

dioxalate		dicarbonate	
Pu (III)	Pu (VI)	Pu (III)	Pu (VI)
2×10^9	2.9×10^{11}		10^{15}

TABLE 3 : Groundwater composition.

Ca^{++}	7.6×10^{-5} M
Mg^{++}	1.3×10^{-4} M
$Fe^{++} + Fe^{3+}$	1.8×10^{-6} M
Na^+	0.24 M
K^+	2×10^{-4} M
Cl^-	1.8×10^{-4} M
SO_4^{2-}	5.2×10^{-6} M
CO_3^{2-}	3.35×10^{-3} M
pH	8.35
Eh	0.15 V

basis of the very little difference between ionic potentials of PuO_2^{++} and Pu^{3+} $/12\_7$. We have thus decided to assume a similar inversion for Pu (III) and Pu (VI) dicarbonates ; through a simple proportion the stability constant $K_2 = 7x10^{12}$ has been attributed to Pu (III)-dicarbonate complexes. On the basis of similar arguments, the stability constants for mono- and tri-carbonate complexes of Pu (III) have been calculated to be, respectively, $K_1 = 3.6x10^9$ and $K_3 = 1.4x10^{16}$. These values have then been used in the present study; however, it is obvious that an experimental verification of the estimated values would be necessary.

An experimental determination of the stability constant for the carbonate complex of Pu (III) could be useful also to evaluate the same constant for Pu (IV). In fact, by comparing the redox standard potentials of the semicouple Pu(IV)/Pu(III) in a non-complexing medium and in a carbonate solution, it is possible to calculate the ratio of the stability constants for Pu (III) and Pu (IV) carbonate complexes; if one of them is experimentally known, the other can then be calculated.

As regards point II), we have decided to treat that stability constant as a variable parameter, to put in evidence its influence in governing the predominance regions of the different plutonium species; this is illustrated in the next paragraph.

5. RESULTS OF THE STUDY

The alpha factors defined in the first paragraph have been used in the present study to quantify the complexation phenomena affecting plutonium in groundwaters.

A typical groundwater composition has been assumed (Table 3), which corresponds to an aquifer in contact with a clay bed. As already stated, in such a chemical environment, the carbonate concentration governs the plutonium complexation.

We have developed a computer code to calculate alpha factors and to assess the plutonium distribution among the different species and oxidation states; to this aim, the redox potential and the various complexation equilibria are considered by the code. Disproportionation reactions have been neglected, in consideration of the low concentration of plutonium.

As discussed in paragraph 3, the numerical value of the stability constant $K_{(IV)}$ of Pu (IV)- carbonate complexes has not

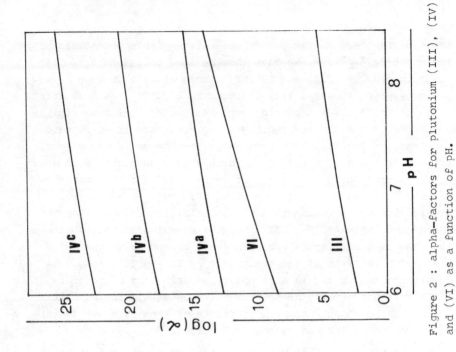

Figure 2 : alpha-factors for plutonium (III), (IV) and (VI) as a function of pH.

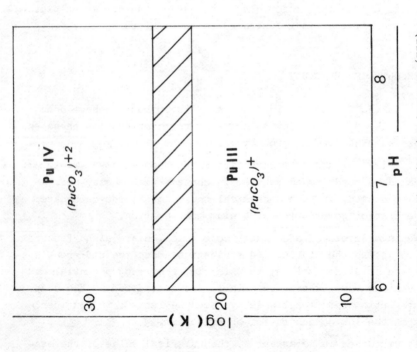

Figure 1 : plutonium distribution between (III) and (IV) states as a function of K(IV) and pH at Eh = 0.15.

yet received a satisfactory quantitative evaluation; therefore, we have treated it as a variable parameter. Such a procedure should permit to identify possible critical ranges of values, which further should be experimentally tested.

The results of this parametric study are shown in figures 1 to 4.

In Fig. 1, the value of $K_{(IV)}$ has been varied from 10^{+10} to 10^{+40}, in the pH range 6-8.5, at Eh = 0.15 (reducing conditions). For each couple of values of $K_{(IV)}$ - pH, the plutonium distribution among different chemical forms and oxidation states has been calculated. The lower part of Fig. 1 represents the region where Pu (III) is the predominant form; the upper part shows the region of Pu (IV) predominance, while the intermediate dashed zone represents a coexistence region, where Pu (III) and Pu (IV) carbonate concentrations differ by no more than one order of magnitude. It may be seen that, also in these reducing conditions, if $K_{(IV)}$ is larger than 10^{25}, nearly all the plutonium results stabilized at the (IV) state, as carbonate complex of the form $\angle PuCO_3(OH)_n \angle^{2-n}$. In the Pu (III) region the complex $\angle PuCO_3 \angle^+$ is predominant, with only a small fraction of Pu^{3+} ion at pH$\sim$6, while for pH$>$8 dicarbonate complexes become appreciable.

In Fig. 2, alpha-factors are reported as a function of pH; curves III and VI show the behaviour of alpha-factors respectively for Pu (III) and Pu (VI), calculated from the stability constant values previously discussed; curves IV a, IV b and IV c show alpha-factor behaviour for Pu (IV) when the stability constant $K_{(IV)}$ is set at the values of 10^{20}, 10^{25} and 10^{30} respectively. The curves of alpha-factors for Pu (III) and Pu (IV) exhibit a very similar slope : this justifies the horizontal slope of the coexistence region in Fig. 1, since in the whole pH range the ratio between the alpha-factors of the two species remains constant. Probably this does not correspond to the real situation, where more OH^- groups are held in Pu (IV) complex than in Pu (III) complex, so that the dashed region of Fig. 1 should exhibit a positive slope.

In Fig. 3, the same calculations of Fig. 1 have been repeated under oxidizing conditions (Eh = 0.6); in this case, the competition occurs between Pu (IV) and Pu (VI), the Pu (III) concentration being now negligible. For low values of the stability constant $K_{(IV)}$, Pu (VI) becomes the predominant species, mainly under the form of the anionic complex $\angle PuO_2 \cdot CO_3 \cdot OH \angle^-$; the dashed

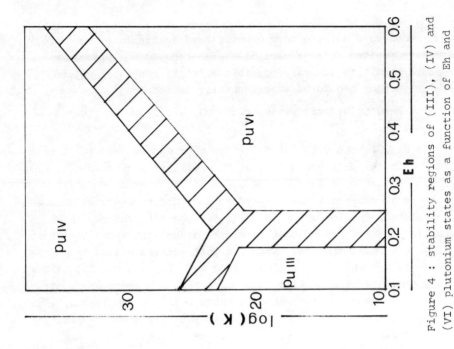

Figure 4 : stability regions of (III), (IV) and (VI) plutonium states as a function of Eh and K(IV), at pH = 8.

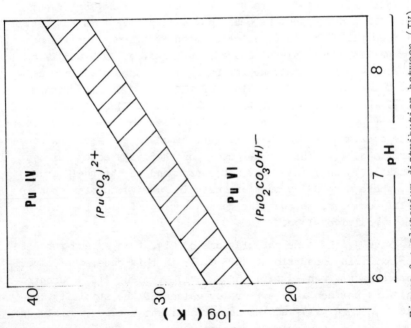

Figure 3 : plutonium distribution between (IV) and (VI) states, as a function of K(IV) and pH at Eh = 0.6.

zone shows a region in which both valence states coexist, at concentrations of the same order of magnitude; in the upper region
Pu (IV)   is the only valence state present. It is worth noting
that in this case the coexistence region limits are not independent on pH : this may be explained on the basis of OH^- group
presence in the Pu (VI) complex.

From figures 1 and 3, it is apparent that, if $K_{(IV)}$ would exceed
a given value (which depends on Eh and pH), the Pu (IV)-carbonate
complex would be the only important species, both in reducing as
well as in oxidizing conditions. This fact is better illustrated
in Fig. 4, where, at pH = 8 the regions of stability of the different plutonium valence states are shown as a function both of
the Eh and $K_{(IV)}$.

It may be seen that for sufficiently high values of $K_{(IV)}$, Pu (IV)
can predominate over the entire range of redox potentials; for
lower values, either Pu (III) or Pu (VI) are the species of major
importance, as a function of the reducing or oxidizing conditions
of the system. In fact, when Eh increases, the concentration of
Pu (III) decreases, being gradually substituted by PuO_2^+ and
PuO_2^{++} ; it is worth noting that, if no complexing anions were
present in the solution, the ion PuO_2^+ would be the most important
species.

When the concentration of PuO_2^{++} approaches the value

$$\frac{\alpha_{(III)}}{\alpha_{(IV)}} \; \llbracket Pu^{3+} \rrbracket$$

the coexistence region (dashed zone) is reached, at Eh between
0.18 to 0.25. For Eh $\approx$ 0.22, pH = 8 and log $K_{(IV)} \approx$ 22 a triple
point is observed, in which plutonium is distributed nearly equally among three different valence states.

A small fraction of PuO_2^+ is present in the coexistence region
between Pu (III) and (VI). A smaller concentration of carbonate
ions would result in the appearance of a Pu (V) region.

It is worth noting that, in order to evaluate the real importance of pH changes in governing the plutonium speciation, the precise form of the complexes should be known, in particular for
what concerns the OH^- groups held in every carbonate complex.
In our opinion, values of $K_{(IV)}$ larger than $\sim 10^{30}$ are not realistic; therefore, in oxidizing environment Pu (VI) carbonates
should be the actual predominating species, while a certain com-
petion could exist between Pu (III) and Pu (IV) for reducing as
well as for intermediate Eh values.

CONCLUSIONS

Plutonium can be present in natural waters in different chemical states, as a function of the redox potential, pH and chemical composition of the medium. In the different valence states, it can form a great variety of soluble complexes with the inorganic anions which are commonly present in groundwaters; among them, the most important ones are the carbonates, because of the great stability of their complexes; quantitative information on the corresponding equilibrium constants is however scarse and unreliable.

On the other hand, an accurate knowledge of the stability of such kind of complexes would be of prime importance to permit a reliable modelling of plutonium migration through porous underground media.

The most important equilibrium constants for which accurate and reliable values are not yet available are those of Pu (III) and Pu (IV) with carbonates; for the former, a value has been preliminarily assessed, by extrapolation of stability constants of similar oxalate complexes; for the latter, a parametric study has been performed, in order to identify possible critical ranges of values. In fact, the value of 10^{47}, proposed by some authors, would lead to unacceptable conclusions.

The parametric study we have performed shows that critical values for this constant could fall in the range of $10^{25} - 10^{30}$; should it be greater, only Pu (IV)-carbonate complexes would be the predominant species, independently on Eh and pH conditions; on the contrary, for lower values, Pu (III)-carbonates and Pu (VI)-carbonates would govern the system, as a function of the redox potential.

As a conclusion, accurate experimental measurements of these stability constants are recognized to be of fundamental importance, in order to allow the description of the chemical system which could migrate underground.

REFERENCES

[1] GIRARDI F., BERTOZZI G., D'ALESSANDRO M.
"Long-term risk assessment of radioactive waste disposal
in geological formations" EUR 5902 EN (1978)

[2] BONDIETTI E.A. and SWEETON F.H.
"Transuranic Speciation in the Environment" in Proc.
Symp. on Transuranics in Terrestrial and Aquatic Environ-
ments. NVO - 178 (1977)

[3] AMES L.L., DHANPAT RAI, and SERNE R.J.
"A rewiev of actinide-sediment reactions with an annota-
ted bibliography"
USERDA Rep., Battelle Pacific Northwest Lab. BNWL-1983,
(1976)

[4] POLZER W.L.
"Solubility of Plutonium in Soil/Water Environments"
In : Proceedings of the Rocky Flats Symposium on Safety
in Plutonium Handling Facilities. April 13-16, 1971.
USAEC CONF-710401. pp. 411-429.

[5] SILVER G.L.
"Disproportionation of Pu (IV) and Pu (V) ions in the
presence of Pu (VI)" MLM-1744 (1970)

[6] SILVER G.L.
"Plutonium disproportionation reactions : Some unresolved
problems" MLM-1807 (1971)

[7] SILVER G.L.
"Plutonium in Natural Waters" MLM-1870.
Mound Laboratory, Miamisburg, Ohio (1971)

[8] SILVER G.L.
"Aqueous Plutonium Chemistry : Some minor problems"
MLM-1871 (1972)

[9] RAI D. and SERNE R.J.
"Plutonium Activities in Soil Solutions and the Stabili-
ty and Formation of Selected Minerals"
J. Environ. Qual. 6 (1), (1977)

[10] POLZER W.L. and MINER F.J.
"Plutonium and Americium behaviour in the soil/water
environment. II. The effect of selected chemical and

physical characteristics of aqueous Pu and Am on their
sorption by soils". Proc. of an Actinide-sediment reactions working meeting : Seattle Feb. 10-11 - BNWL-2117
(1976)

[11] ANDELMAN J.B. and ROZZELL T.C.
"Plutonium in the Water Environment"
In : Radionuclides in the Environment. Adv. In. Chem.,
N° 93, American Chemical Society (1970) pp. 118-137.

[12] MOSKVIN A.L. and GELMAN A.D.
J. Inorg. Chem. USSR (Engl. Translation) Vol. III N°4
(198-216) (1958)

[13] DAHLMAN R.C., BONDIETTI E.A. and EYMAN L.D.
"Biological Pathways and Chemical Behaviour of Plutonium
and Other Actinides in the Environment" In : Actinide
in the Environment. Amer. Chem. Soc. Symposium Series
N° 35.American Chemical Society, (1976)

[14] DAVIDOV Ju.P.
"Sur la nature des colloides des éléments radioactifs"
Radiokhimiya 9 (99-109) (1967)

[15] KELLER C.
"The chemistry of transuranic Elements"
Verlag Chemie GmbH, Weinheim Germany

[16] CLEVELAND J.M.
"The Chemistry of Plutonium" Gordon and Breach Science
Publishers, Inc., New York (1970)

[17] DHANPAT RAI and SERNE R.J.
"Solution species of Pu-239 in oxidizing environment :
I. Polymeric Pu (IV)"; PNL-SA-6994 (1978)

[18] CASADIO S. and ORLANDINI F.
"Double Layer effect on the electrode kinetic reduction
of Pu (IV) carbonate complexes in aqueous solution"
J. Inorg. Nucl. Chem. 34 (3845- 3849) (1972)

[19] GEL'MAN A.D., MOSKVIN et al.
"Complex compounds of transuranium elements";
Consultants Bureau (N.Y.) (1962)

Discussion

P.P. de REGGE, Belgium

In the pH region from 6 to 8 in your diagrams, HCO_3^- will be the predominant form for carbonates in solution. Was this taken into consideration for your calculations ?

A. SALTELLI, CEC

Yes ; for each value of pH our calculation code solves the equilibrium of carbonates and indicates the concentrations of CO_3^{--}.

F.P. SARGENT, Canada

Should the effects of radiation on plutonium species migration be considered ?

A. SALTELLI, CEC

No, we did not consider this because of the low plutonium concentration ($\sim 10^{-8}$M).

G. BERTOZZI, CEC

I wish to add something in reply to Dr. Sargent. In our study we have assumed that the leaching groundwater is flowing continuously, so that the leached plutonium is transported far away from the radiation source as soon as it leaves the glass.

R.H. BECK, Switzerland

You have repeatedly pointed out, that an accurate knowledge of the stability of complexes such as plutonium carbonates are of primary importance to permit a reliable modelling of plutonium migration in groundwater. Do you at Ispra have a comprehensive experimental programme to continue your efforts in this most important field of research ?

F. GIRARDI, CEC

We have under way a four years experimental programme (1977-1980) of which Avogadro has shown a few preliminary results for what concerns chemical speciation of actinides. A second four years programme is under preparation, which partially overlaps the present one (1980-1983). In this programme, under the project "Protective barriers against radionuclides migration" we have included all the experimental actions which we consider necessary for hazard evaluation studies, including the ones which were briefly indicated here. I also would like to add that in all our hazard evaluation studies we assume that engineered barriers will contain radioactivity for a few hundreds years at least. The external radiation effects after such long periods will of course be low.

J. ROCHON, France

Les constantes de stabilité des complexes que vous avez utilisées, ont-elles été déterminées dans des conditions semblables aux conditions naturelles, notamment force ionique, pH, Eh, pCO_2 ?

A. SALTELLI, CEC

No, all the constants that we used in our work were
determined by several authors in the laboratory, generally in conditions far from the natural ones.

"MIGRATION MODELLING OF DIFFERENT PLUTONIUM
CHEMICAL FORMS THROUGH POROUS MEDIA"

A. Saltelli
Commission of the European Communities

ABSTRACT

Two solutions of the migration equations are described. The first
relates to the transport equations for the decay chain Am 243 →
Pu 239 → U 235. Numerical integration was performed in this case
by a simulation code written in CSMP III language and Plutonium is
considered to be all in the same chemical form. The second case relates to the problem of Pu speciation and migration. The decay chain
Pu 240 → U 236 is considered and numerical integration is performed
by a modified version of Bo code COLUMN.
Pseudo first order reactions are supposed to act between Pu states
to maintain equilibrium during the migration.

1. INTRODUCTION

Radionuclides migration through a geologic medium along a preferential pathway can be described by equation

$$D_i \frac{\partial^2 c_i}{\partial x^2} - V \frac{\partial c_i}{\partial x} - \sum_j S_{ji} c_i + \sum_k S_{ik} c_k = K_i \frac{\partial c_i}{\partial t} \qquad (1)$$

where :
$$\begin{aligned}
c_i &= \text{concentration of species } i \\
D_i &= \text{diffusion coefficient} \\
V &= \text{flowing water velocity} \\
K_i &= \text{ion exchange constant} \\
S_{ji} &= \text{rate constant for the homogeneous first order} \\
&\quad \text{reaction } j \rightarrow i \\
S_{ik} &= \text{rate constant for the homogeneous first order} \\
&\quad \text{reaction } k \rightarrow i
\end{aligned}$$

In turn ion exchange can be expressed in terms of the so called "sorption constant" K_{Di} :

$$K_i = 1 + \frac{\rho}{\varepsilon} K_{Di} \qquad (2)$$

where ρ is soil density and ε its porosity. Boundary conditions for eq. 1 are :

$$\begin{aligned}
c_i (x,t) \Big|_{t=0} &= 0 \qquad\qquad c_i (x,t) \Big|_{x=\infty} = 0 \\
c_i (x,t) \Big|_{x=0} &= c_i^0 \, u (t) \qquad \text{with } u(t) = \begin{cases} 1 \text{ if } t \leqslant T \\ 0 \text{ if } t > T \end{cases}
\end{aligned}$$

where T is leach time, i.e. the time needed for water to deplete the matrix containing the radionuclides.

Analytical solutions of equation (1) are generally known [1, 2] but their evaluation, even by computer, may result tedious. The number of terms involved is high and some series evaluation are required to avoid computer overflow and underflow [3]. For these reasons numerical solutions are generally preferred. In the present paper we made use of two calculation codes to perform numerical integration of eq. (1). The first is a simulation code written in CSMP III language, that we applied to the treatment of the typical decay chain Am 243 → Pu 239 → U 235. The second is the Bo code [4] that we adapted with some slight modifications to the problem of Pu speciation and migration.

2. RESULTS AND DISCUSSION

Our simulation code SIMUL is very practical in routine calculation, when complex physico–chemical phenomena are not involved. If
only the decay is considered and first order or pseudo first order
reactions are neglected, the stability conditions for eq. (1) can
be written $\boxed{5}$:

a) $\dfrac{D_i}{\Delta x} > V$

b) $\dfrac{K_i}{\Delta t} \quad \dfrac{2\,D_i}{(\Delta x)^2} - \dfrac{V}{\Delta x} + K_i\,\lambda_{ji}$

where Δx and Δt are space and time increment respectively and
λ_{ji} is the decay constant for the reaction $i \rightarrow j$. Decay constant
appears multiplied by the exchange constant as decay acts on both
adsorbed and solution phases.
It is easy to see that using a fixed grid of coordinates, it is
practically impossible to perform the integration unless very small
space and time increments are adopted. This problem, arising from
the mixed hyperbolic–parabolic nature of this equations is generally
solved by recurring to running coordinates. In this work the problem
has been bypassed by a mathematical artifice which consisted in avoiding that convection and diffusion could act simultaneously. So if
DI and VI are the input diffusion coefficient and water velocity we
set at the integration step K

$\quad$ D = 0
$\quad$ V = 2VI
and at the step K + 1

$\quad$ D = 2DI
$\quad$ V = 0

and so on. Both diffusion and convection are in fact time proportional in our case. Stability conditions reduce now to :

$$\dfrac{K_i}{\Delta t} > \dfrac{4\,D_i}{(\Delta x)^2} + K_i\,\lambda_{ji}$$

that leads to Δx and Δt values much easier to handle. An example
of application of SIMUL code to the decay chain Am 243 $\longrightarrow$ Pu 239 $\longrightarrow$
U 235 is given in figures 1 to 4. The listing of the program, con-
taining also input data, is given below.

INPUT DATA AND PRINT OF THE DATA

```
* LISTING DEL PROGRAMMA DI CALCOLO IN CSMP III PER LA
* SOLUZIONE NUMERICA DELL' EQUAZIONE DI TRASPORTO

INITIAL

NOSORT

*          CALCOLI ESEGUITI IN KM/SECOLI

D=3.0E-06
ROEPSI=1.
DX=0.0025
X=0.1
LEACHT=5.
V=1.

* DATI AM243

KEQ=2000.
NO=100.
LAMBDA=8.72E-03
TABLE NI(1-50)=50*0.0

* DATI PU239

NOS=0.0
KEQU=3000.
LAMBDS=2.84E-03
TABLE NIS(1-50)=50*0.0

* DATI U235

NOR=0.0
LAMBDR=9.72E-08
KEQR=1000.
TABLE NIR(1-50)=50*0.0

K=1.+ROEPSI*KEQ
C=1.+ROEPSI*KEQU
P=1.+ROEPSI*KEQR

FIXED KU
FIXED LL
FIXED I,M,MM,MMM
FIXED NN
KU=0
FIXED II
II=1
MM=50

WRITE(6,10) D
   10 FORMAT(1H1,29X,'COEFF. DI DIFF. (IN KM2/SECOLO) = ',E10.3)
WRITE(6,11) ROEPSI
   11 FORMAT(1H0,29X,'RAPPORTO RO SU EPSILON (IN G/ML) = ',E10.3)
WRITE(6,12) DX
```

```
   12 FORMAT(1H0,29X,'DELTA X  (IN KM) = ',E10.3)
WRITE(6,13) X
   13 FORMAT(1H0,29X,'LUNGHEZZA  DEL PERCORSO (IN KM) = ',E10.3)
WRITE(6,14) V
   14 FORMAT(1H0,29X,'VELOCITA  DELL ACQUA (INKM/SECOLO) = ',E10.3)
WRITE(6,15) N0
   15 FORMAT(1H0,29X,'FLUSSO INIZIALE DEL AM243 (IN % /ANNO) = ',E10.3)
WRITE(6,16) KEQ
   16 FORMAT(1H0,29X,'COSTANTE DI SORPTION AM243 (IN ML/G) = ',E10.3)
WRITE(6,17) LAMBDA
   17 FORMAT(1H0,29X,'COST. DECADIMENTO AM243 (IN SECOLI-1)  = ',E10.3)
WRITE(6,18) N0S
   18 FORMAT(1H0,29X,'FLUSSO INIZIALE DEL PU239(IN % /ANNO) = ',E10.3)
WRITE(6,19) KEQU
   19 FORMAT(1H0,29X,'COSTANTE  DI SORPTION PU239(IN ML/G) = ',E10.3)
WRITE(6,20) LAMBDS
   20 FORMAT(1H0,29X,'COST. DECADIMENTO PU239 (IN SECOLI-1) = ',E10.3)
WRITE(6,30) N0R
   30 FORMAT(1H0,29X,'FLUSSO INIZIALE DEL U235 (IN % /ANNO) = ',E10.3)
WRITE(6,31)LAMBDR
   31 FORMAT(1H0,29X,'COST. DECADIMENTO U235  (IN SECOLI-1) = ',E10.3)
WRITE(6,32) KEQR
   32 FORMAT(1H0,29X,'COSTANTE  DI SORPTION U235 (IN ML/G) = ',E10.3)
WRITE(6,21) LEACHT
   21 FORMAT(1H0,29X,'TEMPO DI LISCIVIAZIONE (IN SECOLI) = ',E10.3,////)

WRITE(6,22)
   22 FORMAT(1H1,45X,'INIZIO OUTPUT CSMP III',/////)
N0=N0/10000.
N0S=N0S/10000.
N0R=N0R/10000.

DYNAMIC

NOSORT
KU=KU+1
IF(KU.EQ.50) WRITE(6,100) TIME
IF(KU.EQ.50) KU=0
  100 FORMAT(20X,'TIME = ',F10.3)
NN=MM-1
X1=STEP(0.0)
X2=STEP(LEACHT)
X3=X1-X2
IF(TIME.EQ.LEACHT) X3=1
INPN  =N0*X3
INPS  =N0S*X3
INPR  =N0R*X3
IF(II.EQ.1) D=0.
IF(II.EQ.1) V=2.
IF(II.EQ.1) GO TO 101
D=6.E-06
V=0.
  101 CONTINUE
II=-II

NDT(1)=((D/K)*(N(2)-2.*N(1)+INPN  )/(DX)**2)-((V/K)*(N(1)-INPN)/DX) ...
-LAMBDA*N(1)
DO 1 I=2,NN
    NDT(I)=((D/K)*(N(I+1)-2.*N(I)+N(I-1))/(DX)**2)-((V/K)*(N(I)- ...
```

- 169 -

```
    N(I-1))/DX)-LAMBDA*N(I)
      1 CONTINUE
NDT(MM)=((D/K)*(0.0-2.*N(MM)+N(NN))/(DX)**2)-((V/K)*(N(MM)-N(NN))/ .
DX)-LAMBDA*N(MM)
N=INTGRL(NI,NDT,50)


NDS(1)=((D/C)*(S(2)-2.*S(1)+INPS   )/(DX)**2)-((V/C)*(S(1)-INPS)/DX)
-LAMBDS*S(1)+LAMBDA*N(1)
DO 2 I=2,NN
      NDS(I)=((D/C)*(S(I+1)-2.*S(I)+S(I-1))/(DX)**2)-((V/C)*(S(I)- .
S(I-1))/DX)-LAMBDS*S(I)+LAMBDA*N(I)
      2 CONTINUE
NDS(MM)=((D/C)*(0.0-2.*S(MM)+S(NN))/(DX)**2)-((V/C)*(S(MM)-S(NN))/ .
DX)-LAMBDS*S(MM)+LAMBDA*N(MM)
S=INTGRL(NIS,NDS,50)
NDR(1)=((D/P)*(R(2)-2.*R(1)+INPR   )/(DX)**2)-((V/P)*(R(1)-INPR)/DX)
-LAMBDR*R(1)+LAMBDS*S(1)
DO 3 I=2,NN
      NDR(I)=((D/P)*(R(I+1)-2.*R(I)+R(I-1))/(DX)**2)-((V/P)*(R(I)- .
R(I-1))/DX)-LAMBDR*R(I)+LAMBDS*S(I)
      3 CONTINUE
NDR(MM)=((D/P)*(0.0-2.*R(MM)+R(NN))/(DX)**2)-((V/P)*(R(MM)-R(NN))/ .
DX)-LAMBDR*R(MM)+LAMBDS*S(MM)
R=INTGRL(NIR,NDR,50)


JAM243=N(10)*10000.
JPU239=S(10)*10000.
JU235=R(10)*10000.
MAM243=N(40)*10000.
MPU239=S(40)*10000.
MU235=R(40)*10000.

TERMINAL
METHOD RECT
OUTPUT MAM243,MPU239,MU235
LABEL CURVA DI LISCIVIAZIONE DELLA CATENA AM243-PU239-
LABEL U235 - MAM243,MPU239,MU235 SONO I FLUSSI IN
LABEL USCITA ( IN CY/ANNO )
OUTPUT N(1-50)
PAGE CONTOUR
LABEL AM243
OUTPUT S(1-50)
PAGE CONTOUR
LABEL PU239
OUTPUT R(1-50)
PAGE CONTOUR
LABEL U235
OUTPUT TIME,MAM243,MPU239,MU235
PAGE XYPLOT
LABEL CURVA DI LISCIVIAZIONE DELLA CATENA AM243-PU239-
LABEL U235 - MAM243,MPU239,MU235 SONO I FLUSSI IN
LABEL USCITA ( IN CY/ANNO )
TIMER FINTIM=500.,OUTDEL=10.,DELT=0.2
END
STOP
```

Concentration pulses at the exit of the soil column ($L = 0.1$ Km) are shown in fig. 1. It can be noted that the arrival times of the radionuclides obey the equation :

$$t_{arrival} = \frac{L}{V} \; K_i \qquad\qquad (3)$$

The diffusion effect of course increases with the permanence time of the radionuclides in the column of soil. Fig.s 2 to 4 show a planimetric shape of the functions c_i (x,t).

To handle more complex phenomena, and in particular the problem of Pu speciation and migration, we have adapted with some modifications the code proposed by Bo $\underline{/4\underline{7}}$. A subroutine (PLUTO) which assesses the distribution of Pu among its oxidation states and complexes, once the physico-chemical properties of the solution are specified (Eh, pH, analytical concentration of complexing agents such as car-bonate etc.), has been inserted at the beginning of the code. It has to be stressed that many of the data used by the subroutine PLUTO are only a rough estimate of unknown constants, and that the problem of Pu speciation in aqueous carbonate solution has not been yet solved $\underline{/6\underline{7}}$.

At the exit from the vitreous matrix Pu distributes among its valence states which are removed by the flowing water. Polymer formation was neglected in view of the chemical features of the solution cho-sen in this example (analytical concentration of Pu 10^{-8}M, carbona-tes concentration 10^{-4}M). Water velocity in this case was supposed to be 10 mt/year and leach time to be 500 years. It has to be poin-ted out that the velocity of removal of the different Pu valence states from the neighbourhood of the waste containers is not the same, since the velocity of each component is governed by its ex-change constant, i.e.

$$V_i = \frac{V}{K_i}$$

We have so chosen the chemical features of the leaching water in order to have the same order of magnitude for the concentrations of Pu III, V and VI. Whith these data input (see tab. 1) Pu IV con-centration is negligible. Further the equilibria of the various va-lence states with Pu IV have been neglected.

$K_{PuCO_3(OH)_n^{1-n}}$	= 1000	T_{leach}	= 500 y
$K_{PuO_2^+}$	= 2000	D	= 0.03 mt^2/y (for all the radionuclides)
$K_{PuO_2CO_3OH^-}$	= 100	pH	= 8.0
$K_{UO_2^{2+}}$	= 100	Eh	= 0.23
V	= 10 mt/y	C_{Pu}	= 10^{-8} M
L	= 100 mt	C_{carb}	= 0.10^{-5} M

TABLE 1 - Input data

In the present example the Pu species available for migration are $PuCO_3(OH)_n^{1-n}$, PuO_2^+, and $PuO_2CO_3OH^-$. Exchange constants are given in table 1. The Pu isotope taken into account is Pu 240, and the decay reaction Pu 240 $\longrightarrow$ U 236 is considered. In order to take into proper account the equilibria between the various valence states during the migration, the following pseudo first order reactions have been considered :

$$Pu\ III \quad \underset{S_{3,5}}{\overset{S_{5,3}}{\rightleftarrows}} \quad Pu\ V$$

$$Pu\ V \quad \underset{S_{5,6}}{\overset{S_{6,5}}{\rightleftarrows}} \quad Pu\ VI$$

Problems at this stage may arise because the reaction constants must simultaneously :

a) - obey stability conditions,
b) - be high enough to allow equilibrium during migration,
c) - the S_{ij}/S_{ji} ratio must be equal to the equilibrium value.

For a given reaction i $\longrightarrow$ j the reaction time is given by

$$t = \frac{0.693}{S_{ji}} \tag{4}$$

and, to respect condition b), it must be much smaller than the permanence time of the radionuclide in the soil column. This condition must be verified without violating the stability conditions for S_{ji} :

$$s_{ji} < \frac{K_i}{\Delta t} - \frac{2^{D_i}}{(\Delta x)^2}$$

In the Bo example it was sufficient, to obtain equilibrium, that
the reaction time for a given radionuclide were about 1/100 of
the permanence time. In our case this ratio must be at least
1/10000. This happens because our exchange constants are very high
and the difference between the highest and the lowest values of the
constants large. In our version of the Bo code the program itself
chooses the S values once the value of the equilibrium ratio and
space and time increments is assigned. In fig. 5 a ratio 1/1000
was used. It can be seen that local peaks are present when equilibrium is not respected. In fig. 6 a ratio $1/10^{+5}$ was used, and
equilibrium is maintained throughout the arrival period. With this
value there are not local peaks and the concentration profiles
show broad shapes between the arrival time of the fastest radionuclide and that of the slowest one. It has to be noted that the presence of a chemical species of Pu with an high sorption constant
will affect also the arrival times and the concentration levels of
the other species with lower constants. This can constitute a self-
induced barrier to Pu migration in geologic formations. As in carbonate containing waters Pu III and VI complexes are likely to be
anionic, the presence of an appreciable amount of PuO_2^+ and/or of
some cationic complex of Pu IV may be of extreme importance in determining the overall rate of arrival at the exit of the disposal
of the radionuclide and its concentration level.
If on the other hand the equilibrium allows only submicro concentrations of cationic species, the migrating pulses of anionic complexes
will not be delayed nor flattened. Experimental studies about the
charge and the stability constants of the carbonate complexes of
Pu in the range of Ph and Eh likely to be found in groundwaters are
therefore of primary importance.

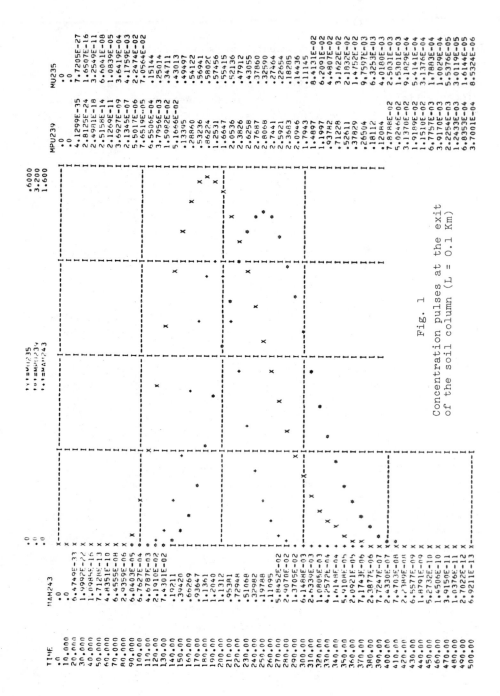

Fig. 1

Concentration pulses at the exit
of the soil column (L = 0.1 Km)

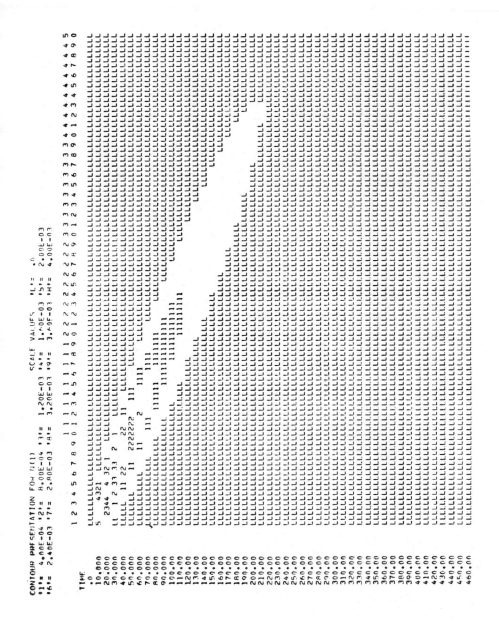

Fig. 2

Planimetric shape of the
function $C_{Am243}(x,t)$

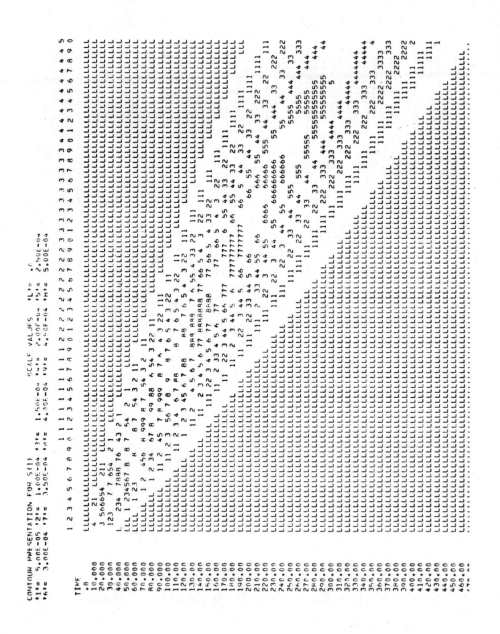

Fig. 3

Planimetric shape of the function $C_{Pu239}(x,t)$

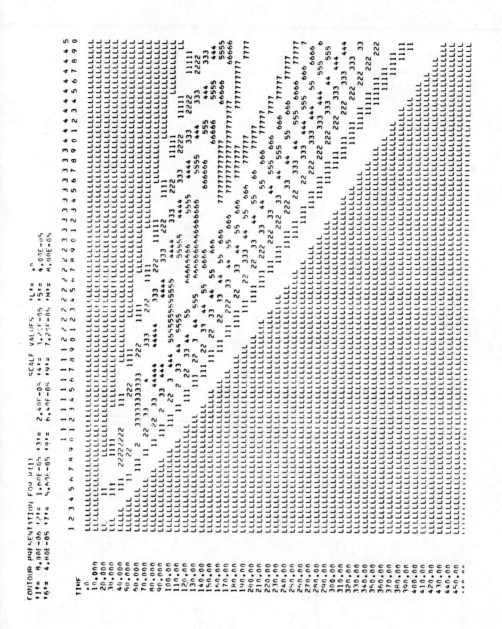

Fig. 4

Planimetric shape of the function $C_{U235}(x,t)$

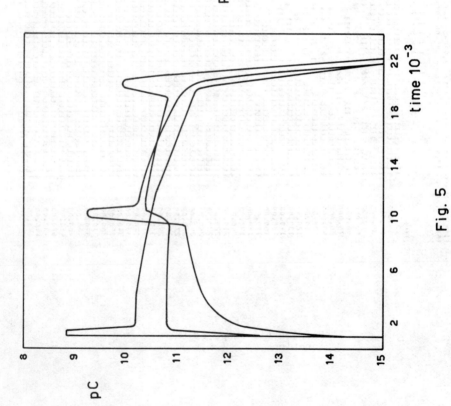

$R \simeq 10^{-3}$

Fig. 5

Concentration pulses for Pu species. R is the ratio between reaction times and permanence times. Peaks at time=1000,10000 and 20000 relate to Pu VI , III and V respectively.

Fig. 5

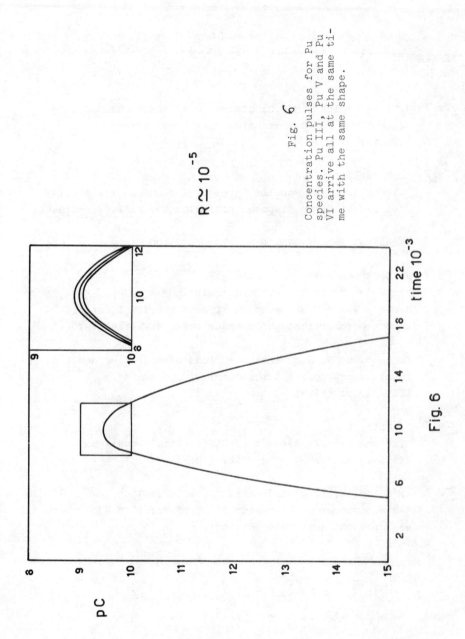

$R \simeq 10^{-5}$

Fig. 6

Fig. 6

Concentration pulses for Pu
species. Pu III, Pu V and Pu
VI arrive all at the same time with the same shape.

REFERENCES

/ 1 / H.C. Burkholder, M.O.Cloninger, D.A. Baker and G. Jansen
"Incentives for partitioning high level waste"
BNWL 1872 (1972)

/ 2 / A. Saltelli
"Alcune note sui modelli matematici connessi all'analisi
dei rischi di un disposal geologico per rifiuti radioatti-
vi"
Technical Report CNEN - RT/PROT. (77) 20

/ 3 / H. C. Burkholder, G. Jansen
"Actinide Chain Migration Calculation"
Nuclear Waste Management and Transportation Quarterly
Report October through December 1974, BNWL 1899-1975

/ 4 / P. Bo "COLUMN, Numerical solutions of migration equations
involving various phisico-chemical processes"
RISO Report (1978)

/ 5 / L. Collatz
"The numerical treatment of differential equations"
III Ed., Spinger Verlag Edit. (1966)

/ 6 / A. Saltelli, A. Avogadro and G. Bertozzi
"An Assessment of Plutonium Chemical Forms in Groundwater"
Work presented at this workshop.

LE PROGRAMME RADIOPROTECTION DE LA COMMISSION :
ACTIVITES DE RECHERCHE EN MATIERES DE TRANSFERT DES RADIONUCLEIDES
ET D'EVALUATION DES RISQUES D'IRRADIATION

C. Myttenaere
Responsable du Secteur "Comportement et contrôle
des radionucléides dans l'environnement"
du Programme Radioprotection de la CCE (Bruxelles)

RESUME

Description des activités de recherche conduites dans le cadre du Programme
Radioprotection de la Commission des Communautés Européennes.
Dans cet exposé, l'accent est mis sur toute recherche présentant un intérêt
évident en matière de stockage des déchets radioactifs.

ABSTRACT

Description of the research activities which are conducted in the framework of the
Radiation Protection Programme of the Commission of the European Communities.
The scope of the paper is to stress each scientific work which is related to the
storage of radioactive wastes.

1.   LE PROGRAMME RADIOPROTECTION

Le rôle du Programme Radioprotection de la Commission consiste en l'acquisition des données nécessaires à la compréhension et au contrôle des risques d'irradiation engendrés par les utilisations diverses de l'énergie nucléaire.

L'évaluation des conséquences biologiques et écologiques de ces diverses activités a, par conséquent, pour objectif d'assurer une protection appropriée de l'homme et de l'environnement, dans tous les cas où des dommages inadmissibles pourraient leur être infligés.

Le Programme comporte six activités ou secteurs principaux étroitement intégrés :

- la dosimétrie des rayonnements et son interprétation,
- le comportement et le contrôle des radionucléides dans l'environnement,
- les effets somatiques à court terme des rayonnements ionisants,
- les effets somatiques à long terme des rayonnements ionisants,
- les effets génétiques des rayonnements ionisants,
- l'évaluation des risques d'irradiation.

Deux de ces secteurs intéressent plus spécialement le problème du stockage des déchets radioactifs :

- le comportement et le contrôle des radionucléides dans l'environnement,
- l'évaluation des risques d'irradiation.

Ces activités de recherche s'adressent par conséquent plus spécialement aux barrières rétention des radionucléides en sol profond et transfert des radionucléides dans l'environnement / 1 /.

2.   CONTAMINATION DE L'ENVIRONNEMENT SUITE AU STOCKAGE DES DECHETS

Plusieurs événements peuvent causer la dispersion de la radioactivité enfouie en couches géologiques profondes : phénomènes naturels à déroulement rapide (activité volcanique, météorites) ou lent (érosion, changement de climat), modifications de la géologie locale (effets de température ou impact mécanique sur les roches), interventions humaines (forages, explosions nucléaires...).

La réinsertion directe des radionucléides dans la biosphère par le truchement de telles manifestations semble peu probable et seule la possibilité d'un transfert via l'eau (suite à un mouvement sismique par exemple) a été retenue jusqu'à présent dans le cadre du stockage en couches géologiques.
Tenant compte des connaissances actuelles en matière de physico-chimie des radionucléides et dans le cadre d'un lessivage de la matrice de verre (1000 ans après dépôt), les premiers radionucléides susceptibles de contaminer le milieu seraient les Tc-99 et l'I-129 et ce 200 ans après le contact avec les eaux souterraines / 2 /. A plus longue échéance ($10^4 - 10^6$ ans) les radionucléides suivants devraient être considérés comme critiques : Np-237, Sc-79, Cs-135, Zr-90, Pd-107 et Sn-126.

Les radionucléides susceptibles de contaminer les eaux de boisson seraient Np-237, Ra-226, Tc-99 et I-129. Ceux-ci peuvent aussi entrer dans la chaine alimentaire et, au travers des maillons de celle-ci, atteindre l'homme.

Toutes ces études sont toutefois entâchées "d'incertitudes" en ce qui concerne les radionucléides à retenir et la valeur des paramètres considérés dans les divers modèles ; ces valeurs demandant confirmation pour les conditions de milieu envisagées.

D'autre part, les études théoriques faites par le CCR /¯1_/ dans le cadre du programme stockage des déchets montrent de plus que les déchets α de basse activité (bituminized α-bearing low level waste : BIP) posent également des problèmes en ce qui concerne la contamination de l'environnement. Dans le cadre de l'étude (formation saline) et des hypothèses de départ, une pollution de l'environnement n'atteindrait des niveaux significatifs que si les pertes se manifestent endéans les 10^4 ans après dépôt ($\neq$ actinides). Dans le cas d'une contamination accidentelle du milieu après 10^5 ans, le radionucléide le plus dangereux s'avérerait être le Ra-226.

Ce rapport porte également l'attention sur diverses approximations faites lors de cette étude et sur certaines incertitudes qui subsistent (méconnaissance des phénomènes de resuspension, variation des transferts dans le temps, interactions diverses, extrapolation des résultats à divers sites...).

Des travaux suédois récents ont également été consacrés au problème du stockage des déchets en roches précambriennes /¯3_/. Selon ceux-ci, la décomposition lente de la matière pourrait augmenter légèrement le niveau de la radioactivité du milieu après plusieurs milliers d'années. Cette légère augmentation étant due surtout aux nucléides suivants : NP-237, Tc-99, Ra-226, U-233, Cs-135 et I-129. Les doses reçues seraient toutefois dans ce cas plus faibles que les doses maximales admissibles fixées par l'ICRP (eau de boisson puisée à proximité du lieu de stockage) pour des personnes vivant à proximité des installations nucléaires.

Bien que nous disposions à l'heure actuelle de pas mal de données sur le transfert de certains de ces radionucléides, la plupart des facteurs de transfert utilisés ont été obtenus en conditions contrôlées ou suite à des circonstances différentes (retombées radioactives) qui ne permettent pas l'extrapolation de ces résultats aux conditions propres au stockage des déchets !

Le programme radioprotection actuel s'efforce de combler ces lacunes en considérant des "situations réelles" et en exploitant des données obtenues sur le terrain.

3. RECHERCHES EFFECTUEES DANS LE CADRE DU PROGRAMME ACTUEL

Les figures suivantes schématisent les diverses recherches conduites dans le domaine du transfert et de l'évaluation des risques.

Les principaux buts poursuivis par ces recherches sont :

- L'étude des mécanismes physiques responsables de la diffusion et de la dispersion des radionucléides,
- L'influence de la physicochimie du milieu ainsi que du vieillissement sur les transferts,
- la nature et l'importance des interactions entre radionucléides et polluants associés,
- le calcul des demi-vies biologiques,
- L'établissement de modèles de dispersion propres aux différents modes de contamination du milieu.

Dans le cadre de ces recherches, nous nous sommes efforcés de maintenir une collaboration étroite avec le programme stockage actions directe et indirecte.

4. RECHERCHES PREVUES DANS LE CADRE DU PROGRAMME 1980-1984

- Secteur : Comportement et contrôle des radionucléides dans l'environnement.

Dans le cadre de ce programme, la priorité sera donnée aux radionucléides et aux voies de transfert dans l'environnement susceptibles de jouer un rôle important dans les programmes d'énergie nucléaire au cours des prochaines

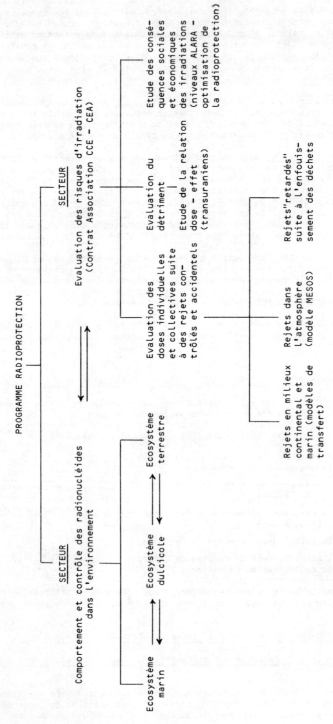

PROGRAMME RADIOPROTECTION

SECTEUR
Comportement et contrôle des radionucléides dans l'environnement

SECTEUR
Evaluation des risques d'irradiation
(Contrat Association CCE - CEA)

Ecosystème marin ⟷ Ecosystème dulcicole ⟷ Ecosystème terrestre

Evaluation des doses individuelles et collectives suite à des rejets contrôlés et accidentels

Evaluation du détriment

Etude de la relation dose - effet (transuraniens)

Etude des conséquences sociales et économiques des irradiations (niveaux ALARA - optimisation de la radioprotection)

Rejets en milieux continental et marin (modèles de transfert)

Rejets dans l'atmosphère (modèle MESOS)

Rejets "retardés" suite à l'enfouissement des déchets

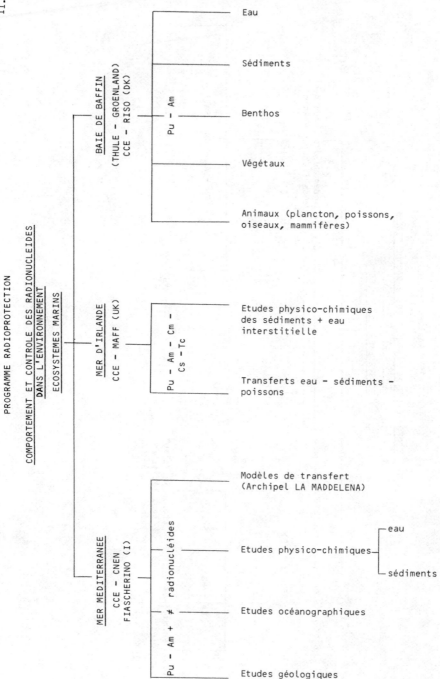

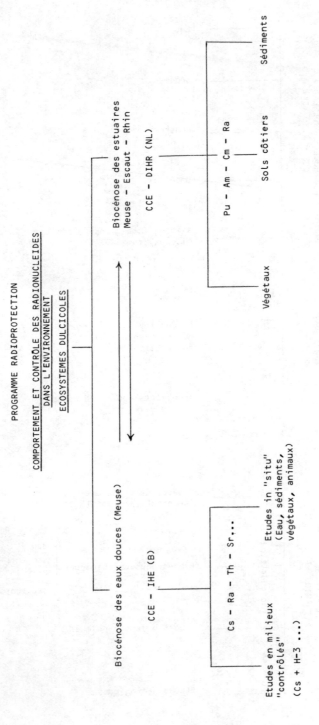

PROGRAMME RADIOPROTECTION

COMPORTEMENT ET CONTROLE DES RADIONUCLEIDES
DANS L'ENVIRONNEMENT
ECOSYSTEMES DULCICOLES

Biocénose des eaux douces (Meuse)

CCE - IHE (B)

Cs - Ra - Th - Sr...

Etudes en milieux
"contrôlés"
(Cs + H-3 ....)

Etudes in "situ"
(Eau, sédiments,
végétaux, animaux)

Biocénose des estuaires
Meuse - Escaut - Rhin

CCE - DIHR (NL)

Pu - Am - Cm - Ra

Végétaux

Sols côtiers

Sédiments

III.

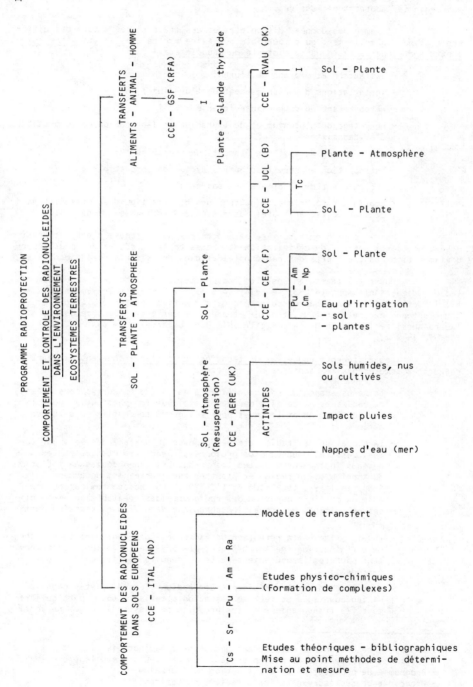

décennies, ou par suite de l'introduction éventuelle de substances radioactives dans l'environnement à partir d'autres sources. Le programme détaillé sera élaboré en tenant compte d'autres programmes de la Communauté (1) s'adressant à la sécurité nucléaire et la protection de l'environnement, ce qui permettra par conséquent de maintenir les contacts nécessaires.

L'examen des données disponibles et des travaux prévus lors des différents stades du traitement du combustible nucléaire montre que <u>les activités suivantes sont importantes dans le contexte du programme</u> :

- extraction et broyage de l'uranium ;

- installations d'enrichissement de l'uranium ;

- retraitement du combustible irradié ;

- recyclage du plutonium et de l'uranium et fabrication de combustibles d'ocydes mixtes ;

- introduction de systèmes élaborés de réacteurs ;

- introduction éventuelle d'autres cycles de combustible ;

- déclassement de réacteurs nucléaires ;

- gestion, y compris élimination, des déchets liquides, gazeux et solides, produits par toutes les activités mentionnées ci-dessus.

Une attention particulière sera accordée aux méthodes d'estimation du niveau de contamination, de délimitation des zones contaminées, et de réduction ou d'élimination des transferts des radionucléides dans des situations accidentelles.

Outre les transuraniens, les radionucléides qui semblent actuellement les plus importants sont le H-3, le C-14, le S-35, le Kr-85, le Tc-99, le Ru-106, l'I-129 et l'I-131, ainsi que certains produits d'activation (Mn-54, Co-60) et radioisotopes naturels (radium, thorium et produits dérivés). Dans ce cadre, il faudra aussi tenir compte de la toxicité chimique de certains de ces nucléides (Tc-99, I-129).

<u>Les différents processus de transfert dans l'environnement</u> exigeant des recherches complémentaires sont résumés ci-dessous :

- la resuspension de radionucléides à partir de la surface des océans, de sédiments et de sols européens typiques (en particulier pour le Np, le Pu, l'Am et le Cm et pour les produits de fission à longue période) ;

- la migration des radionucléides déposés en surface des sols agricoles, vers les couches plus profondes, l'eau, les plantes et les animaux (notamment pour les transuraniens, les produits de filiation du thorium et du radium et d'autres radionucléides, notamment le S-35, le Tc-99, le Ru-106 et l'I-129). La radiocontamination des animaux par incorporation des radionucléides préalablement métabolisés et suite à une exposition chronique demande une attention particulière ;

- la migration et la rétention de radionucléides dans certaines roches et certains types de sol typiques des pays de la Communauté (notamment pour les transuraniens et les produits de fission à longue période) ;

- la rétention par les sédiments des radionucléides rejetés dans le milieu aquatique et leur mobilisation ultérieure (plus particulièrement les transuraniens et les produits de fission à longue période) ;

---

(1) - Programme de gestion et de stockage des déchets radioactifs
    - Programme de recyclage du plutonium dans les réacteurs à eau légère
    - Programme de prospection et d'extraction de l'uranium
    - Programme de déclassement d'installations nucléaires
    - Programme concernant la sécurité des réacteurs à eau légère

- la distribution confinée et le comportement des radionucléides à longue période (par exemple le C-14, le Tc-99, l'I-129), en particulier en ce qui concerne leur échange entre différentes composantes de l'environnement (par exemple échanges entre les écosystèmes aquatiques et terrestre) ;

- l'absorption par des espèces aquatiques de radionucléides particuliers (par exemple le Tc-99), pour lesquels des informations complémentaires sont nécessaires ;

- l'étude des effets de synergie éventuels entre radionucléides et polluants conventionnels rejetés dans l'environnement, spécialement en ce qui concerne l'absorption des radionucléides dans la chaîne alimentaire ;

- l'échange de C-14 et HTO entre l'atmosphère et l'environnement terrestre ;

- la dispersion atmosphérique et les processus de dépôt dans les zones urbaines.

- <u>Secteur</u> : <u>Evaluation des risques d'irradiation</u>

On envisagera trois groupes de problèmes dans ce secteur :

Le premier concerne <u>l'évaluation des doses individuelles et collectives</u> résultant d'émissions normales et de rejets accidentels de substances radioactives. Cette évaluation des doses doit s'appuyer sur les données obtenues par l'étude des mouvements des radionucléides dans l'environnement et devrait conduire à une meilleure détermination de la répartition des doses parmi la population et de l'ampleur de la dose collective, compte tenu du fond naturel. Des modèles s'imposent également pour toutes les voies d'accès éventuelles à l'homme et à son environnement et ces modèles doivent porter sur la totalité du cycle nucléaire.

En ce qui concerne l'optimalisation de la radioprotection qui est actuellement préconisée, il importe de tenir compte également des risques provoqués par les activités humaines qui font appel aux rayonnements ionisants ou qui ont une influence sur l'irradiation, par exemple celles qui concernent les applications médicales et la radioactivité technologiquement renforcée. Le programme comprendra des phases successives en ce qui concerne l'identification des points à étudier, l'évaluation des doses reçues par les travailleurs et par le public et l'étude des mesures de protection possible et de leur coût.

Le second problème concerne la recherche méthodologique de <u>l'évaluation des dommages</u>. On utilisera les données recueillies grâce à la recherche expérimentale et épidémiologique décrite dans les secteurs ad hoc du programme. Il conviendra de retenir deux groupes de problèmes. D'abord, ceux qui concernent l'évaluation du dommage causé par une irradiation moyenne et élevée, applicable en cas d'accident. Ensuite, les problèmes relatifs aux faibles doses, qui présentent un intérêt tout particulier pour toutes les personnes exposées aux radiations sur le plan professionnel.

Le troisième problème concerne <u>l'évaluation des conséquences économiques et sociales</u> de l'irradiation. Il s'agit d'un nouveau thème qui devrait être développé si l'on veut mettre au point des directives pour l'optimalisation de la radioprotection qui soient fondées sur l'obtention des "niveaux d'irradiation les plus faibles auxquels on puisse raisonnablement prétendre (ALARA)", dans les conditions en vigueur en Europe.

Références bibliographiques

/ 1 /   GIRARDI, F., BERTOZZI, G. et D'ALESSANDRO, M. : Long-term risk assessment of radioactive waste disposal in geological formations, C.E.C., Nuclear Science and Technology, J.R.C., ISPRA, EUR 5902 EN, 42 pp (1978).

/ 2 /   HEINONEN, O. et MIETTINEN, J.K., Introduction to radiological aspects of radioactive waste disposal, Nordic Society for radiation Research and Technology, 6th Conference, Hirtshals, DK, August 28-31 (1978).

/ 3 /   XXX., Handling of spent nuclear fuel and final storage of vitrified high level reprocessing waste, Safety analysis, KBS Report 125 pp (1977).

/ 4 /   XXX., Rapports annuels Programme Radioprotection (1976 - 1977).

GENERAL DISCUSSION   DISCUSSION GÉNÉRALE

D. RAI, United States

Most of the papers presented in this session indicate that we must know the chemical form of the various radionuclides present in the water, if we want to understand their behaviour. It was also pointed out by various people that such data are not now readily available. The best approach is probably a thermo-dynamic one. However, there are some problems with the thermo-dynamic approach, because the data available at the present time are not very accurate. For the elements which exhibit more than one oxidation state, another problem is that the species in solution are going to be controlled by the oxidation-reduction potential of the soil system. There is a controversy whether we can measure reliably the oxidation-reduction potentials of soil media and what those measured values mean. We cannot translate the existing process data with radical Eh-pH conditions to the expected ranges of natural systems. I wonder if somebody has comments or suggestions about the best approach to this problem.

F.P. SARGENT, Canada

It seems to me that we have to do more experiments in situ; then, measurements would be comparable. The problem is how to relate what we do in the lab, to the in situ conditions. I much prefer to do things in the lab and I have no experience of doing experiments in situ. Perhaps, that is why I am suggesting that we do more in situ experiments, because it always seems easier in somebody else's system.

D. RAI, United States

I definitely agree. If we want to take a look at a particular site, we must study it in situ. However, some of these in situ experiments may not give us a good clue to the mechanism involved, because the number of variables involved in in situ experiments is so large that if we try to understand the mechanisms and the behaviour of the elements, we may not get many answers. However, in situ experiments would indicate what might happen in that particular environment.

P. TREMAINE, Canada

It would be a great help if we were to do experiments in soils or deep formations; it would facilitate the computations greatly.

L.R. DOLE, United States

In the United States nuclide migration programme, the focal point for collection of thermo-dynamic data, and its application, mainly using the Helgeson code, is the University of California, Berkeley. The principle researcher is Dr. John Apps, working,

- 191 -

of course, in conjunction with Don Rai's group at Battelle Northwest. I would encourage you to exchange data and to address these
researchers directly and have some rapport with them. These wor-
kers should be the main contact points in the United States programme. In order to operate the Helgeson code at the University of
California, they have collected most of the pertinent thermodynamic data from the United States.

F. GERA, NEA

          In relation to geologic disposal of waste, we are really
interested in long-term and large-scale problems. A good deal of
information could be obtained by looking at natural in situ experiments that are performed in geological systems. Of course, we
do not have many actinides in natural geological systems, but there
are uranium, thorium and their daughter products that can be considered in particular situations as geochemical analogues of the
higher actinides. By observing what happens around uranium and
thorium deposits, it is possible to learn a lot about the likely
behaviour of neptunium, plutonium and americium in similar geological environments.

          Fall-out is another large-scale experiment that has been
provided free to us. In this particular case, we know that there
is a good deal of information on plutonium and a little less on the
behaviour of neptunium, americium and curium in soils. I think
that these data are very important, because they refer to the large-
scale behaviour of the elements. We could never perform such large-
scale experiment under coutrolled conditions.

C. MYTTENAERE, CEC

          Je voudrais reprendre la remarque qui a été faite par le
Dr. Sargent au sujet des expériences in situ et en laboratoire, et
je m'empresse de dire que je suis biologiste et non pas géologue ;
j'ai été frappé au cours de ces exposés par le fait que les expériences en laboratoire ne reflètent certainement pas, du point de
vue biologique, les conditions qui se passent à de très grandes
profondeurs, et que par conséquent, si l'on veut pouvoir extrapoler
ce résultat aux conditions naturelles, il faudrait que l'on tra-
vaille en conditions beaucoup plus stériles qu'on ne l'a fait jusqu'à présent.

D. RAI, United States

          Certainly, the deep underground conditions are quite dif-
ferent from the conditions at the surface. In our programme in the
United States, we are measuring sorption coefficients under natural
environment and oxydizing conditions. We are also completing studies under anoxic conditions, in which the oxygen is completely
excluded. So at least, these lab data compare somewhat to that
natural situation.

P.C. LEVEQUE, France

          Nous sommes étonnés, je dis nous, parce que nous avons eu
déjà quelques échanges d'idées à ce sujet, de voir que dans ce programme que nous a donné M. Myttenaere, personne n'a parlé jusqu'à
maintenant de ce qui a été fait précédemment. Puisque je suis
français, je ne parlais pas des travaux qui ont été faits en France
mais enfin il y a eu les travaux de notre collègue M. Baetsle qui
ont ouvert une voie considérable, peut-être pas sur des produits
à période longue ; il y a eu des travaux à la Trisaia en liaison

avec le CNEN et l'EURATOM ; je pense qu'il y a eu également des
symposias qui ont   été tenus sur Oklo, le nom n'a pas été prononcé
jusqu'à maintenant. Nous avons eu au Congrès International de Madrid
en septembre dernier,"la géologie de l'ingénieur", un exposé particulièrement passionant de ce qui a été réalisé en Suède entre une
équipe américaine et une équipe suédoise dans les granites. On ne
parle pas des travaux français qui ont été plus modestes dans certains cas, mais enfin, certains d'entre nous ne comprennent pas
très bien qu'il n'y ait pas de liaison entre les travaux antérieurs
qui ont coûté un certain nombre d'unités de compte, les travaux qui
sont faits dans d'autres cadres et dans d'autres coordonnées géo-
graphiques de la Communauté, alors que finalement ceci intéresse
à peu près tout le monde ; le plutonium n'est ni suédois, ni amé-
ricain, ni italien, ni français, ni autre ; je pense que ce sont
des produits que l'on retrouvera quelle que soit la carte et l'iden-
tité du produit ; par conséquent, il me paraît quelque peu curieux
qu'il n'y ait pas, sur le plan aussi bien de la jonction adminis-
trative que financière que des études in situ une liaison, une os-
mose un petit peu plus évidente, un petit peu plus efficace entre
tous ces travaux. Je pense notamment que dans le cas de ce qui a
été exposé par M. Myttenaere, et de ce qui est réalisé dans certains centres, il n'y a non plus cette liaison ; c'est pourquoi, ce qu'a demandé le Dr. Sargent, son collègue canadien, me
paraît tout-à-fait dans une ligne de conduite efficace, probablement assez économique, car on ne ferait pas plusieurs fois certains
travaux.

C. MYTTENAERE, CEC

        Je voudrais répondre à ce qui vient d'être dit, mais je
me limiterai dans cette réponse au programme biologie et radioprotection. Il est bien clair que ce programme s'adresse à la biosphère et ne s'intéresse pas au problème de la géosphère. Je pense
qu'il y a deux éléments de réponse à donner à votre intervention.
Le premier est celui-ci : tous nos résultats et si j'avais
voulu vous donner le compte rendu de nos résultats antérieurs
acquis dans le cadre de programmes précédents, j'aurais dû pouvoir
disposer non seulement de 20 minutes mais d'une demi-journée ; mais
tous ces résultats précédents ont fait l'objet de publications,
d'abord, publications de la part de nos contractants et puis
publications de la Communauté sous forme de rapports annuels. Ces
publications sont à la disposition de tous. En ce qui concerne
ces contacts entre contractants et non-contractants apparte-
nant aux différents pays de la Communauté, je crois que le cadre
dans lequel nous évoluons nous permet justement de provoquer ces
contacts et c'est ce que nous faisons au niveau de notre programme
en organisant régulièrement ce que nous appelons des "workshops"
comme celui-ci, des groupes de travail, des réunions de réflexion
et c'est ce que nous faisons quand nous sommes invités comme obser-
vateurs dans des réunions qui sont organisées par l'AIEA, l'OCDE
ou par d'autres organismes internationaux.

P. COHEN, France

        Est-il possible d'avoir quelque information complémen-
taire sur deux expériences que vous avez citées : l'une concernant
le plutonium qui s'est promené de 30 kilomètres ; est-ce que vous
pouvez nous indiquer quelques données plus précises ? Deuxièmement,
en ce qui concerne les travaux de M. Saas sur le vieillissement du
plutonium dans le sol ; est-ce que sur ces deux sujets il est possible d'avoir oralement des informations ou bien être renvoyé à
des documents publiés ?

C. MYTTENAERE, CEC

A la première question, je voudrais répondre que ce contrat avec le Centre de Risø, au Danemark, vient d'être signé à peine et que par conséquent, nous ne disposons pas, dans le cadre de ce contrat, de données expérimentales. Nous allons participer aux frais d'une nouvelle expédition cette année à Tule, au Groënland, et les récoltes qui seront faites aussi bien de sédiments que de flore et de faune marine feront naturellement l'objet d'une analyse. Ce que je vous ai cité (tantôt), cette dispersion à plus de 30 kilomètres du point d'impact, c'est un résultat qui a été observé par le Dr. Aarkrog et ses collègues au cours d'expérimentations antérieures à notre passation de contrat. En ce qui concerne maintenant la seconde question, je voudrais passer la parole au Dr. Saas puisqu'il est dans la salle et qu'il est certainement l'homme le plus indiqué pour y répondre.

A. SAAS, France

En ce qui concerne le vieillissement du plutonium dans le sol et son transfert à la plante, il est communément admis que le facteur de transfert sol-plante pour les végétaux cultivés est de l'ordre de 10^{-5}. Après sept cultures successives de haricots sur un même sol qui a été contaminé une seule fois au départ, il y a 5 ans, nous en sommes actuellement à un facteur 10^{-3}, c'est-à-dire au fur et à mesure que nous répétons des cultures sur ce même sol, donc que le plutonium vieillit, il y a augmentation du taux de transfert et ce taux de transfert se situe, pour un certain nombre de différences, à la fois en ce qui concerne les formes qui sont assimilées et leur localisation ; c'est-à-dire qu'il y a un vieillissement très important qui est dû à la micro-flore du sol qui est capable de minéraliser ce plutonium et nous observons une diminution du transfert au niveau racinaire, une augmentation du transfert parallèle par le système tige-feuille et, au bout de cinq cultures, passage au fruit vert et au fruit sec. Actuellement, en ce qui concerne ces mêmes transferts, nous avons en expérimentation un système multiple concernant plutonium, americium, curium et neptunium ; nous observons également ce vieillissement, mais un phénomène encore plus important qui est l'effet synergie entre ces quatre radionucléides sur le plan oxydo-réduction, c'est-à-dire que l'apport de neptunium, par exemple, favorise le passage du plutonium au niveau des haricots, qui apparaît au niveau des fruits dès la deuxième culture et cette augmentation se chiffre entre un facteur 2 et 5 selon le type de végétal, et selon la partie du végétal. Par ailleurs, nous avons mis en évidence également, en ce qui concerne ces problèmes sol-plante, toute l'importance des eaux d'irrigation. Il apparaît actuellement que dans les eaux superficielles naturelles telles que le Rhône, la Loire, la Seine ou la Moselle, les transuraniens sont transportés à longue distance à l'état soluble ou pseudo-soluble. J'ai amené ici un certain nombre de photos montrant la migration de ces radioéléments au niveau de ces eaux naturelles et actuellement nous avons pu mettre en évidence la présence de ces complexes solubles dans les eaux du Rhône et parallèlement, nous avons une répartition uniforme pratiquement au niveau des zones cultivées de cette région de l'ordre de 30 picocuries par kilo en plutonium, c'est-à-dire qu'il n'y a aucun gradient de dispersion qui normalement apparaît quant on a des formes hydrauliques, des formes insolubles ou colloïdales ; il y a donc une répartition qui est vraiment une répartition de formes solubles.

D. RAI, United States

I might add that long-term experiments are also being carried out at Battelle Pacific Northwest Laboratories, and they have been going on for about 7 to 8 years. Long-term uptake of plutonium and neptunium is studied in lysimeters, in which the mass balance of the material which is left in the columns is performed.

I have a question for Dr. Saas. It appears to me that there are many uptake factors of actinides by plants which have been reported in the literature. These uptake factors from soil to plant vary widely. It would seem to me that the uptake by plants would merely depend upon how much soluble element is present in the soil. Therefore, the factors found in one particular soil or material may not be applicable in general, because uptake is going to be a function of the soil at that point. I wonder if Dr. Saas would like to comment on what kind of reliability one can expect in uptake factors.

A. SAAS, France

En ce qui concerne le problème sol proprement dit, nous avons choisi exprès un sol pour lequel les taux de transfert étaient considérés comme nuls, c'est-à-dire des sols calcaires où effectivement le pH est tel que nous avons une immobilisation importante. Ensuite, nous sommes en train de faire une étude sur 16 types de sols différents, c'est-à-dire nous représentons pratiquement l'ensemble du territoire français ; les taux de passage sont à peu près du même ordre de grandeur en ce qui concerne leur augmentation, ce qui ne veut pas dire que nous aurons un même taux de transfert pour un type de sol donné mais le plus important, c'est que chaque type de sol, en vieillissant augmente ce taux de transfert. Autrement dit, quel que soit le taux de transfert initial qui est propre à chaque sol, nous sommes certains maintenant que ce taux de transfert dans le temps va augmenter, et donc, il n'est pas question de tenir simplement compte d'un type de sol ou d'un autre, d'une eau d'irrigation ou d'une autre, le problème le plus important est que, en vieillissant, la microflore du sol ou les bactéries ou autre micro-organisme de l'eau sont capables de solubiliser ce plutonium et ces transuraniens et nous obtiendrons une augmentation des taux de transfert. Je pense que c'est là le plus important ; ce n'est pas tellement d'avoir un sol acide où les taux de transfert arriveront jusqu'à 10^{-2} et un sol calcaire à 10^{-3}. Ce qui est important, c'est qu'on aura une augmentation au moins d'un facteur 100, voire peut-être 1000 à l'échelle de quelques années.

C. MYTTENAERE, CEC

Je voudrais ajouter à ce que vient de dire M. Saas, que la grande différence entre une couche géologique et un sol, c'est que dans un sol de surface il y a une microflore et une microfaune qui jouent un rôle extrêmement important dans la mobilisation et dans la modification des formes chimiques des éléments et des radioéléments ; c'est la raison pour laquelle du moins on le suspecte, ce plutonium dans les fonds marins est allé se balader à plus de 30 kilomètres du point d'impact. Il y a eu là au niveau des sédiments marins une action de la flore microbienne qui a transformé ce plutonium - ne me demandez pas sous forme de quel complexe : je crois que le mot "complexe" masque mon ignorance à ce sujet - et qui a fait que ce plutonium plus soluble et moins retenu par les sédiments s'est retrouvé à une telle distance.

M. KHALANSKI, France

Je voudrais rappeler à M. Myttenaere que nous disposons d'un certain nombre de données sur la contamination en plutonium des sédiments de la mer d'Irlande et d'autre part de données françaises qui concernent les parages de l'usine de retraitement de

la Hague (Manche). Il est clair que l'on retrouve du plutonium
bien au-delà des 30 kilomètres autour des rejets de ces deux installations de retraitement. La chose est connue, je pense que l'é-
tude à laquelle vous faites allusion dans votre programme en mer
d'Irlande va confirmer ce fait. Par contre, et c'est le point le
plus important, on observe dans les deux cas un gradient très net
à partir du point de rejet, contrairement à ce qu'a observé
M. Saas, dans un cas où il y avait effectivement une solubilisation de plutonium. Donc, je crois que dans l'état actuel de nos
connaissances, et en tout cas de l'expérience européenne, on peut
conclure qu'en milieu marin le plutonium présente un comportement
bien classique et que le compartiment de stockage est bien le sédiment. Il est possible que l'on trouve du plutonium dans les organismes très loin du point de rejet, mais cela représente actuellement des quantités négligeables, à mon avis.

<u>G. MATTHESS</u>, Federal Republic of Germany

        I wonder if the transport of sediments on the sea floor
has been taken into account in the Baffin Bay and in the Irish Sea
studies. If plutonium is fixed on the sediments, and the sediments
move, spreading of plutonium will take place of course.

<u>C. MYTTENEARE</u>, CEC

        Ce que vous venez de dire est très exact et dans le cadre
du contrat que nous avons avec Lowestoft on s'est aperçu que pour
des raisons apparemment inconnues, il y avait des déplacements ac-
crus de radionucléides et ceux-ci ont été causés probablement par
des brassages, par des mouvements, des remises en suspension du
sédiment dans l'eau et un transport physique et non pas un transport biologique.

<u>P.P. DE REGGE</u>, Belgium

        Je voudrais demander au Dr. Saas s'il a des données sur
la complexation du plutonium par des acides umiques ou des débris
de matières organiques présents dans le sol. Et, si tel est le cas,
si ces données ont été publiées quelque part.

<u>A. SAAS</u>, France

        En ce qui concerne la complexation par les matières orga-
niques, nous avons un certain nombre de données - elles ne sont
malheureusement pas encore publiées, parce que trop récentes -
elles concernent les composés organiques présents dans les eaux de
surface, en particulier des eaux de fleuves comme la Seine, la
Moselle, le Rhône où nous avons observé une complexation importante
au niveau du sol également, mais en milieu uniquement acide où finalement ces complexes peuvent se former avec une constante de stabilité relativement élevée et particulièrement sur des composés à
faible poids moléculaire de type fulvique, c'est-à-dire des molé-
cules relativement petites, dont le poids moléculaire est inférieur
à 2000 et tout notamment sur un certain nombre de composés en pro-
venance de micro-organismes du sol qui sont essentiellement acida-
minés - d'ailleurs, une publication russe vient de paraître à ce
sujet - et un certain nombre d'acides organiques carboxyliques,
maliques, citriques qui sont d'ailleurs utilisés par ailleurs au
niveau médical humain pour simuler les passages des formes complexes
du plutonium. Je voudrais ajouter quelque chose pour vous montrer
les problèmes qui se posent en milieu marin. Nous avons déposé suc-
cessivement americium, curium, neptunium et plutonium sur des gels
de silice, c'est-à-dire sur des milieux inertes et, sur des gels de

silice avec des résines anioniques, après nous avons fait passer
de l'eau de mer. Toutes les taches pratiquement ont disparu, c'est-
à-dire que les transuraniens ont migré en milieu marin, sauf pour
deux éléments, le neptunium partiellement et le plutonium très peu.
Par contre, americium et curium sont partis et donc sont capables
de migrer en milieu marin. Et donc, il y a en fait, en milieu marin
un double problème, celui des transports des éléments figurés,
c'est-à-dire des sédiments proprement dits, plus des transports qui
vont se faire certainement à l'état soluble ou pseudo-soluble.

D. RAI, United States

        I would like to ask Dr. Saas a question. It is very inte-
resting to note that the uptake of plutonium was increasing conti-
nually over the period of your work. I wonder if you would expect
the same thing, if one would start with a more soluble species
of plutonium in the soil. What was the initial plutonium in your
work?

A. SAAS, France

        Certainement pas, puisque les taux d'augmentation agran-
dis ou constatés pour les végétaux varient pour les diverses formes
de plutonium apportées. Il est certain que cette augmentation paraît la plus probable et la plus importante pour les formes insolubles. C'est-à-dire que pour celles-ci l'absorption racinaire est
nulle au départ et celle-ci croît en fonction de la mise en solution. En revanche, pour des formes solubles apportées par l'eau d'irrigation, nous avons un transfert qui sera relativement stable et il
a pû être estimé, par exemple, au niveau des végétaux qui sont cultivés dans l'aire irriguée par le Rhône où nous observons une relative stabilité. Mais ce qui veut dire aussi que dans ces cas-là
l'apport est pratiquement continu puisque l'irrigation est continue
d'année en année et donc il y a un dépôt au sol qui est très diffi-
ficile à estimer. Ensuite, par contre, le transfert paraît relativement stable, mais il est certain que l'apport de formes différentes au sol va automatiquement produire des taux de transfert
différents, comme cela est le cas pour d'autres radionucléides
tels que le ruthenium par exemple, le césium partiellement, ou
même par exemple surtout pour le technetium ou l'iode. Il est certain qu'au niveau des végétaux ou du sol apporter de l'iodate ou
de l'iodure n'aura pas le même effet sur le taux d'absorption. Par
contre, un sol est capable, en lui apportant un iodure, de trans-
former cet iodure en sept formes différentes par exemple qui sont
l'iodate, le périodate, l'iode élémentaire, l'iode à l'état de
méthyl, etc., et chaque forme de cet iode aura un taux d'absorption différent. C'est-à-dire, il faut lui donner une équation de
tranfert qui sera une exponentielle différente. C'est ce que nous
faisons actuellement.

D. RAI, United States

        If there are no further questions, I would like to close
this session. I wish to thank all the people who have presented
papers and all the speakers for a very stimulating discussion. I
would also like to take this opportunity to invite you to participate in the United States' data bank. We would welcome any kind
of data either published or unpublished.

Session IV
Transport in Various Media

Chairman — Président
Dr. F. GIRARDI
(CEC)

Séance IV
Transport dans différents milieux

IN SITU PLUTONIUM - TRANSPORT IN GEOMEDIA

A.T.Jakubick

Kernforschungszentrum Karlsruhe GmbH
D 7500 Karlsruhe   F.R. of Germany

1 Introduction

During handling of fuel cycle plutonium the formation of small amounts of
aerosols cannot be entirely avoided.
Most of the aerosols produced are composed of PuO_2, but some of the plutonium released may come from evaporation of $Pu(NO_3)_4$. In the latter case
$Pu(NO_3)_4 \times 5 H_2O$ is the probable chemical form encountered up to 40°C.
Above this temperature it is transformed to PuO_2; this change is completed
at 250°C.

By various operations aerosols of different AMAD+) particle sizes are produced. In a reprocessing plant the size mainly ranges from 0.1 to 1.0 µ.
In fuel element fabrication the particle size distribution peaks between
1 and 5 µ. In research and development installations the usual size varies
from 1 to 2 µ [1].

In routine work plutonium is exclusively released through HEPA++) filters
with a decontamination factor of 10^{-4} to 10^{-5}. A combination of several filters will result in a decontamination factor of 10^{-7}. Experiments indicate
that the highest penetration is due to the 0.3µ diameter particles.

For safety optimization of potential plutonium emittents of the fuel cycle
it is, therefore, necessary to develop a predictive methodology for the
build up and persistancy of this element in soil.

+) AMAD   = Activity Median Aerodynamic Diameter
++) HEPA-filter = High Efficiency Particulate Air Filter

A straight forward approach to this task would be to investigate how plutonium from the worldwide fallout was distributed during the last two and half decades. In the course of atmospheric nuclear tests from 1954 to 1962 about 18.5×10^{15} Bq Pu-239 were injected into the atmosphere [2]. The size of the particles immediately after an explosion ranges from 0.01 to 1000μ [3].

On their way through the stratosphere and upper troposphere the fallout particles undergo a filtration which results in a size range of 0.02 to 0.3 μ in the lower troposphere [3].

The HEPA-filtered PuO_2 released by industrial plutonium handling [4]and the worldwide fallout PuO_2 filtered by transport seem to have comparable properties after having reached the ground level air. The difference mainly consists in a slight shift to larger particles of the maximum of PuO_2 released by industrial operations.

It is possible to consider the migration of fallout plutonium therefore as a long term in situ experiment when adequate methods of evaluation are applied. We can then use the results for direct predictions concerning surface contamination in microconcentration range.

The complex nature of reactions occuring between plutonium-contaminants and geological media will continue to defy precise analysis for some time. In the absence of an exact theory a black box model concept has been employed. The results are interpreted in terms of characteristic plutonium transmission rates through soil barriers. Subsequently soil physical parameters are attributed to these values.

In addition there is the question whether the lab scale models are compatible with the in situ situations.

2 Data base generation
───────────────────────

In the following section we shall consider a variety of in situ plutonium depth distribution profiles. The requirements on suitable sampling were given in previous papers [4]. An overview of the sampling sites is given in Fig.1 and approximate information on the soil granulometry in Fig.2 .

First we investigated three sites in the vicinity of Heidelberg: a sandy, a loamy and a clay-organic soil profile. The sampled layer thickness was 2 cm(Fig.3).

During a second sampling campaign, south of Karlsruhe and Stuttgart,we concentrated specifically on clay soils.The investigations involved soils with (1)illitic, (2)kaolinitic-mixed-layer, (3)montmorillonitic,(4)illite-montmorillonite-chloritic clay mineral assemblies. The sampled layer thickness was 5 cm (Fig.4).

Finally the vicinity of the Gorleben site (east of Braunschweig in Lower Saxony) was sampled.In addition to 5 estimations of the present plutonium background level (to a depth of 25 cm),three plutonium depth distribution profiles were taken (Fig.5). Two of these profiles were in gley soil (sample layer thickness 2 and 5 cm, resp.) and one in podzolic soil(sample layer thickness 2 cm).

The aim of the approach followed here is to obtain the plutonium-flux and the effective barrier retardation from the distribution pattern at the time of sampling which arose from a given source function operating during the time of plutonium release.

We can then set up a plutonium-in-soil balance for each considered layer and time step:

$$m_{t+1} - m_t = j \cdot c_{input} - j \cdot c_{box} \qquad (1)$$

| change of | input | output |
| inventory | flux | flux |

m = amount of plutonium in the box at time t

c_{input} = concentration of plutonium at the input boundary

c_{box} = concentration of plutonium in the box

j = transmission rate

The relation has a simple physical significance: Eq(1) is a continuity condition which says that in a unit volume of geological media the plutonium-sorption and the losses through transport are equal to the plutonium recharge provided by the source. For the uppermost layer the input is given by the known plutonium fallout deposition [4,5]. For the subsequent deeper layers the output flux is taken as input.

As a rule, only transmission values for layers of comparable thickness can be meaningfully interpreted. It therefore follows that the plutonium concentrations c_{box} must vary with box volume in addition to the variation in amount of plutonium recharge. In the case of a cross section taken as unity, one divides only the plutonium inventory m_t by the layer thickness Δz: $c_{box} = m_t / \Delta z$. We can then obtain the plutonium transmission rates independent of the box size for a representative part of the layer considered. This characteristic j is estimated by reconstructing the whole plutonium migration history of the site (Fig.6) by means of the Eq.1 and adjusting the value of j to the flux required for the balance. The validity is checked by comparing the calculated plutonium inventory with the measured amount of contamination in the sampled layers. Thus j is an effective rate averaged over the whole migration time. The histogram of Fig.7 gives the range of plutonium transmission rates typical for middle European climatic conditions.

Basically, the essential objective is the effective retardation R exerted by the soil barriers. Thus we must relate the velocity of soil water percolation (which we assume transports plutonium) to the transmission rate:
$R = v(H_2O)/j$.

To evaluate $v(H_2O)$ we establish a soil-water balance for each site and layer. The precipitation data are routinely recorded by the Meteorological Service; the actual evapotranspiration can be calculated by using a functional relationship (and empirical calibration factors) between atmospheric conditions and soil water content w. Then the soil water recharge G can be calculated for each layer. Herefrom the percolation follows $v(H_2O) = G/w$.

In the following diagram the values of soil barrier plutonium retardation R are plotted (Fig.8). The site specific relevance of the data also can be identified from the picture.

3 Implications for nuclide/geo-media interaction

For purposes of general application it is of particular interest to in - vestigate how retardation changes with soil parameters. Before proceeding further we have therefore listed these parameters in Tab.1.

In our formulae it is assumed that biological activity does not contribute to the migration which is only approximate. A proper adjustment of the effective transmission time $\bar{t} = 1/j$ can take this into consideration.

Table 1

SITE	No	Γ	pH	p	org.C	carbonate %
GOLF LINKS	1	0.287	5.00	0.537	6.92	<2
	2	0.154	4.50	0.572	4.62	<2
	3	0.0980	4.40	0.466	3.13	<2
	4	0.0925	4.40	0.304	1.15	<2
	5	0.0903	4.50	0.294	3.16	<2
	6	0.0826	4.60	0.330	3.41	<2
	7	0.0787	4.60	0.322	2.49	<2
2 cm layer thickness	8	0.0488	4.60	0.330	2.32	<2
	9	0.0459	4.70	0.319	1.53	<2
	10	0.0481	4.70	0.331	2.52	<2
	11	0.0441	4.60	0.321	2.55	<2
	12	0.0393	4.50	0.343	1.03	2
	13	0.0325	4.50	0.349	0.94	2
	14	0.0304	4.50	0.368	2.30	2
	15	0.0352	4.50	0.348	1.14	2
	16	0.0383	4.50	0.327	1.25	2
GAUANGELLOCH MEADOW	1	0.0600	7.07	0.643	3.80	6.1
	2	0.0286	7.02	0.477	3.30	6.2
	3	0.0366	7.05	0.573	2.90	3.2
	4	0.0373	7.09	0.516	1.60	5.0
	5	0.0335	7.09	0.516	1.60	2.2
	6	0.0223	7.12	0.545	1.20	1.9
	7	0.0246	7.18	0.478	1.80	2.0
	8	0.0192	7.23	0.365	0.70	1.9
	9	0.0113	7.23	0.605	0.90	8.0
2 cm layer thickness	10	0.0137	7.28	0.410	1.00	6.7
	11	0.0103	7.31	0.331	0.50	3.1
	12	0.00945	7.32	0.478	0.50	3.8
	13	0.0152	7.36	0.640	0.50	3.2
	14	0.0124	7.34	0.604	0.20	3.0

Γ = surface density of charge in meq/m^2

p = porosity

org.C = organic carbon in %

Table 1

(continued)

SITE	No	Γ	pH	p	org.C	carbonate %
MUSTHOF FARM	1	0.0212	6.31	0.693	6.20	<1
	2	0.0153	6.33	0.368	5.80	<1
	3	0.0161	6.45	0.474	5.30	<1
2 cm thickness	4	0.0136	6.57	0.421	6.30	<1
	5	0.0127	6.62	0.560	4.30	<1
	6	0.00999	6.56	0.567	4.70	<1
	7	0.00830	6.65	0.419	4.70	<1
	8	0.00940	6.75	0.659	4.20	<1
	9	0.00934	6.67	0.535	5.80	<1
HOHENBOL HILL	1	0.0164	5.70	0.612	6.05	0
	2	0.0127	6.30	0.510	5.86	0
5 cm thickness	3	0.00982	6.10	0.620	4.09	0
	4	0.00836	6.10	0.577	3.26	0
	5	0.00682	6.00	0.549	2.61	0
	6	0.00470	6.00	0.510	2.05	0
SCHRAMBERG MEADOW	1	0.0128	5.90	0.623	5.74	0
	2	0.0132	6.10	0.440	3.74	0
5 cm thickness	3	0.0104	6.20	0.531	3.27	0
	4	0.00817	6.60	0.418	2.51	0
	5	0.00410	6.50	0.477	3.19	0
	6	0.00449	6.50	0.464	1.68	2.0
BOEHRINGEN QARRY	1	0.00412	6.80	0.602	5.00	4
	2	0.00394	7.00	0.503	3.55	5
5 cm thickness	3	0.00377	7.10	0.471	3.60	5
	4	0.00324	7.20	0.468	2.59	5
	5	0.00298	7.20	0.338	2.33	4
	6	0.00262	7.20	0.440	1.57	5
SCHOEMBERG ORCHARD	1	0.00653	5.90	0.639	2.01	0
	2	0.00538	5.70	0.505	1.87	0
5 cm thickness	3	0.00525	5.90	0.526	1.84	0
	4	0.00533	6.00	0.502	1.76	0
	5	0.00453	6.10	0.504	1.09	0
	6	0.00332	5.60	0.421	0.85	0

Γ = surface density of charge in meq/m^2

p = porosity

org.C = organic carbon in %

In every theory (even if not explicitly stated) the velocity of a contaminant is found from: $R = 1 + D_{nuclide}$, where $D_{nuclide}$ is the paramount factor since it determines the velocity of the migration [6]. We therefore give the plutonium retardation by soil barriers analogously:

$$R = K_o + K_1 \cdot D_{Pu} \tag{2}$$

The best fit to the data of Tab.1 we obtain with $K_o = 82.02$, $K_1 = 5.34 \times 10^4$. The soil parameters related to D_{Pu} were chosen according to a factor analysis. As the most important factors were identified: ion exchange capacity T, specific surface s, soil equilibrium pH and porosity p. In first approach the effect of the organic carbon content was considered to be sufficiently represented by the ion exchange capacity. The ratio T/s is used as the surface density of charge $\ulcorner$:

$$D_{Pu} \sim \ulcorner \frac{1}{pH} \ p \ . \tag{3}$$

We now find :

$$R \ \sim \ 82 + 5.3 \times 10^4 \ (\ulcorner \frac{1}{pH} \ p \ ) \tag{4}$$

Within the present accuracy of measurements the relationships (2) and (3) are valid for $3.0 \leq T \leq 34.5$ meq/100g; $0.23 \leq s \leq 61.7$ m^2/g; $4.4 \leq pH \leq 7.3$; $0.26 \leq p \leq 0.6$ with a correlation coefficient of 0.94. Further relevant soil parameters than those considered here certainly exist and also the formation of various plutonium species may complicate the plutonium/geo-media interaction. It would need a chemical-analytical approach to identify them.
Near the zero degree coefficient $K_o = 82$ which is a threshold value in eq(4) the relative contribution of neglected factors would increase. Additional corrections for this plutonium-fraction, and for the additional inter - action effects which can be associated with more complicated reactions , are apparently small. They are likely to be comparable with the degree of error in the plutonium-measurement in the field.

Thus in eq(4) a compromise is made between ease of application and accuracy of prediction.
For completeness sake we also note that fallout plutonium consists of high fired PuO_2, whereas the industrial emission would lead to low temperature PuO_2. Due to a serious lack of experimental information it is not clear whether this difference is of significance.

The remarkable feature of (4) is, that on the left side there are only dynamic properties, whereas on the right side there are only properties of the soil barrier. Looking into the expression (3) one finds: the retardation is primarily electrostatic in nature depending on the surface charge of the soil particles.

The reverse of soil equilibrium pH corrects the effect of surface charge according to the soil proton reservoir. This is an oversimplification as the surface density of charge response to pH is different for clay minerals and organic matter than it is for the usually negatively charged quartz grains. One could speculate that the 1/pH term also includes the effect of the redox potential of the soil as normally $Eh \sim 1/pH$.

The porosity can be regarded physically as a small correction term which accounts for the effectiveness of the surface area for sorption; the higher the porosity, the thinner the diffuse double layer adjacent to the solid phase, the more intimate the contact and more probable the sorption.

The retardation did not show significant change with the clay mineral assemblies. It is interesting to note that we could not even prove a clear retardation difference between clay vs. sand soils.

The clay content, of course, lowers the transmission rate but this effect is largely masked by organic matter content of the soil. More generally we conclude (with 99.98% probability) that above an organic carbon content of some 3% the effect of organic matter on retardation becomes overwhelming (Tab.1). In the range investigated, $0.2 < $ org.C $ < 7\%$, a retardation decrease by 3 1/2 due to organic matter content seems likely.

Below the 3% org.C content the correlation of soil parameter and plutonium
retardation becomes 0.97. This indicates again that eq (4) is strictly
heuristic. One should be cautious about drawing conclusions outside of the
investigated range. The statistically analyzed data set is not strictly
normal distributed (because of nonrandom sampling); linear correlation,
therefore, may be premature. However the nonparametric Spearman rank correlation proved eq.4 positively on the 99.99% level.

4 Application to a radioecological site evaluation

For orientation purposes, first, it is useful to give values of the pre -
sent (1976-1978) plutonium background level in Germany. They represent the
baseline for future emissions.
The total Pu-239, 240 and Pu-238 in an open grassland in SW-Germany amounts
to:

	Pu-239,240 Bq/km^2	Pu-238 Bq/km^2	$\dfrac{Pu\text{-}238}{Pu\text{-}239}$
vicinity of Heidelberg			
Heidelberg golf links	3.92×10^7	1.63×10^6	0.0416
Gauangelloch meadow	7.78×10^7	3.21×10^6	0.0413
Mückenheim Musterfarm	6.03×10^7	2.76×10^6	0.0458
south of Karlsruhe/Stuttgart			
Hohenbol hill	7.67×10^7	2.30×10^6	0.0300
Schramberg meadow	7.89×10^7	2.26×10^6	0.0286
Boehringen quarry	7.02×10^7	2.18×10^6	0.0311
Schoemberg orchard	6.69×10^7	2.12×10^6	0.0317
SW-Germany average	6.56×10^7	2.31×10^6	0.0352
S.D.	1.28×10^7	1.24×10^6	0.0121

In the vicinity of the Gorleben site (near Braunschweig, N-Germany) the
open grassland total plutonium content in soil is:

		Pu-239,240 Bq/km^2	Pu-238 Bq/km^2	$\dfrac{Pu\text{-}238}{Pu\text{-}239}$
Waltersdorf	A	3.53×10^7	1.19×10^6	0.0336
	B	4.57×10^7	1.16×10^6	0.0253
	C	4.52×10^7	2.03×10^6	0.0450
	D	3.95×10^7	1.80×10^6	0.0456
	E	4.21×10^7	1.22×10^6	0.0291
Krautze	I	4.88×10^7	1.49×10^6	0.0306
	II	2.36×10^7	1.51×10^6	0.0642
Gorleben site average		3.91×10^7	1.46×10^6	0.0373
S.D.		1.28×10^7	1.24×10^6	0.0138

As a sensitive indicator of local contamination the Pu-238/Pu-239,240 ratio
can be used.

The SW vs.N-Germany plutonium-background averages differ significantly
on the 99% level. The lower N-German background can be attributed to the
latitude dependence of the worldwide fallout input. The decrease in the plutonium background by a factor of 1.7 to 1.8 corresponds to the latidude in-

Table 2

SITE		No	Γ	pH	p	org.C	R
TREBEL		1	0.0871	3.93	0.67	9.45	716.8
		2	0.1361	3.51	0.79	8.36	1262.0
		3	0.4452	3.45	0.68	2.13	3428.0
		4	0.2135	3.42	0.53	1.52	1325.3
		5	0.1740	3.41	0.48	1.43	995.6
		6	0.1330	3.41	0.46	1.62	748.1
		7	0.1154	3.44	0.42	1.20	607.1
		8	0.1329	3.41	0.41	1.34	673.7
		9	0.0879	3.48	0.36	1.23	420.7
		10	0.0428	3.48	0.38	0.84	248.6
		11	0.0277	3.61	0.36	0.81	179.6
	2 cm layer thickness	12	0.0271	3.80	0.43	0.32	198.1
		13	0.0229	3.90	0.48	0.42	190.9
		14	0.0197	4.02	0.44	0.15	164.9
		15	0.0262	4.17	0.38	0.72	179.5
		16	0.0275	4.24	0.42	0.43	241.5
		17	0.0339	4.24	0.45	0.23	238.1
		18	0.0244	4.22	0.45	0.15	190.4
		19	0.0279	4.23	0.43	0.15	
GORLEBEN		1	0.0471	2.89	0.86	8.23	518.3
		2	0.0388	2.79	0.68	4.30	341.8
		3	0.2724	2.94	0.50	2.12	1580.6
		4	0.2456	3.03	0.35	2.15	1022.3
		5	0.2078	3.12	0.45	1.73	1106.6
		6	0.2043	3.20	0.48	1.50	1157.7
		7	0.2738	3.29	0.48	1.24	1527.8
		8	0.2990	3.35	0.43	1.31	1495.9
		9	0.1981	3.38	0.41	1.12	970.5
		10	0.0841	3.47	0.36	1.24	404.9
	2 cm layer thickness	11	0.0392	3.58	0.35	0.92	221.1
		12	0.0677	3.69	0.44	0.36	399.5
		13	0.1783	3.85	0.34	1.31	741.6
		14	0.2302	4.06	0.28	1.62	784.4
		15	0.1843	4.21	0.49	1.24	1071.4
		16	0.1084	4.26	0.48	1.89	646.5
		17	0.0654	4.29	0.45	0.24	395.6
		18	0.0442	4.35	0.37	0.42	250.3
		19	0.0375	4.32	0.43	0.31	247.8

Table 2

(continued)

SITE	No	Γ	pH	p	org.C	R
BRÜNCKENDORF	1	0.1835	4.38	0.72	2.01	1535.0
	2	0.1639	3.93	0.55	1.72	1069.6
	3	0.1515	3.77	0.30	1.42	573.3
	4	0.1065	3.76	0.44	1.53	589.0
	5	0.0633	3.78	0.37	1.20	328.7
	6	0.0455	3.82	0.33	0.95	231.7
	7	0.0365	3.87	0.32	0.82	198.3
	8	0.0285	3.91	0.33	0.46	173.1
	9	0.0237	3.96	0.27	0.21	139.8
2cm layer thickness	10	0.0196	4.04	0.22	0.36	116.7
	11	0.0234	4.12	0.34	0.21	157.3
	12	0.0183	4.14	0.34	0.31	137.9
	13	0.0218	4.15	0.30	0.37	141.3
	14	0.0223	4.14	0.32	0.42	148.0
	15	0.0252	4.17	0.36	0.36	169.4
	16	0.0207	4.15	0.38	0.71	156.3
	17	0.0113	4.14	0.33	0.28	110.3
	18	0.0069	4.14	0.37	0.42	97.0
	19	0.0079	4.15	0.35	0.63	99.5

crease of $4°$ to the north (between $49°$ to $53°N$).

We now have enough experimental data and interconnecting theory to permit evaluations of the potential build up of the plutonium level in the soil at the future Gorleben nuclear site.

Mineralogicaly the soil constituents here are 80-90% quartz, 7-15% feldspar and ca 5% clay minerals. Montmorillonit and illit prevails alternately. We summarize the soil parameters for the site in Tab.2 .

Presently the best value for the plutonium-transmissivity of the ecologically important top 25 cm soil layer for Gorleben follows from the plutonium-depth distribution profiles of Fig.5 evaluated by eq.(1) as:

$$j_{0-25} = 0.82 \text{ cm/year.}$$

The solution of (4) for the estimated soil parameters (Tab.2) gives the retardation and transmission rate for the deeper layers:

$$j_{25-50} = 0.61 \text{ cm/year}$$

$$j_{>50} = 1.35 \text{ cm/year .}$$

We assume the following emission source operation parameters:

isotope composition	reactor Pu with 2% Pu-238
operation period	25 years
maximum local fallout	1.9×10^8 Bq/km^2 . year
average (over 60 km^2) local fallout	6.3×10^7 Bq/km^2 . year .

The calculation using eq.(1) leads to a maximum plutonium accumulation of 4.5×10^8 Bq/km^2 and an average of 1.5×10^8 Bq/km^2 in the upper layer[Fig.9].

The accumulation would persist for some 75 years at the place of maximum concentration and about 68 years at an average area of 60 km^2. Thereafter the 1978 background would be reached.

During downward migration the contamination peak becomes smoothed at about 50 cm. At a depth of 100 cm we already observe a large dispersion.
The ground water table at 10m is reached by the peak of the contamination wave after 780 years.
At this time the plutonium-wave maximum amounts to 8.3×10^7 Bq/km^2 and the average 2.8×10^7 Bq/km^2.
The calculations performed account only for the expected routine release.

5. Conclusions

Finally the following conclusions can be drawn:

1- The present (1976-1978) total soil plutonium background in SW-Germany is 6.6×10^7 Pu-239 Bq/km^2 and 2.3×10^6 Pu-238 Bq/km^2 and in N-Germany 3.9×10^7 Pu-239 Bq/km^2 and 1.5×10^6 Pu-238 Bq/km^2.
This would be the average plutonium baseline level for an open grassland for any future emission.

2- Due to a continuous decrease of the fallout plutonium input source towards the north the plutonium background east of Braunschweig is lower by a factor of 1.7 than around Heidelberg/Karlsruhe.

3- The plutonium background diminishes by downward migration into the soil at a rate of 0 to 3.5 cm/year under middle European climatic conditions.

4- The retardation of plutonium compared to the partially saturated water flow (the barrier effect of the soil) was in 84% of cases below 300 and in 16% of cases below 100.

5-For levels greater than 3% organic carbon the organic matter present in soil is the most effective sorption parameter for plutonium. More gene- rally a correlation between surface density of charge of soil particles and plutonium retardation seems to exist.

6-When using the soil parameters of the future Gorleben nuclear site and the emission parameters of maximum 1.9×10^8 Bq/km^2 . a$_2$(average 6.3×10^7 Bq/km^2 . a) a plutonium accumulation of 4.5×10^8 Bq/km^2 (average 1.5×10^8 Bq/km^2) can be predicted for the upper 25 cm top soil after 25 years operation.

ACKNOWLEDGEMENTS

The author is obliged in particular to G.Gutzeit(for programming), I.Kahl (for sampling and lab-work), G.Hoeland and M.Nesovic (for techni- cal assistance),Mrs.H.Gottschalk(for editorial remarks).The helpful comments of Dr.H.Krause are also gratefully acknowledged.We appreciate the kind cooperation of Dr.Heger,German Central Meteorological Service,Offen- bach and the soil analytical support made available by Prof.G.Müller,Inst. Sedimentary Research,University Heidelberg.The program is sponsored by Project Safety Studies Entsorgung(coordinated by Prof.H.W.Levi,Hahn-Meit- ner-Inst.Berlin).

REFERENCES

1.J.C.Elder,Gonzales,M. and Ettinger H.J."Plutonium aerosols size charac - teristics".Health Phys. 27, 45-53 (1974) .

2.Hardy,Krey and Volchok "Global inventory and distribution of fallout plutonium" Nature 241 (1973) .

3.C.E.Junge "Radioactive Aerosole" Nuclear Radiation in Geophysics Springer Verlag Berlin-Göttingen-Heidelberg (1962)

4.L.H.Ahrens,Ed. Origin and distribution of the elements,Proceedings of the Second Symposium, Paris-UNESCO,May 1977, 775-790,Pergamon Press(1979)

5.C.N.Welsh, Ed.Transuranium Nuclides in the Environment,Proceedings of a Symposium San Francisco 17-21 Nov.1975, 47-62,IAEA,Vienna (1976)

6.M.J.Frissel and P.Poelstra,"Chromatoraphic transport through soils" Plant and Soil XXVI,no.2 (1967)

Figure 1

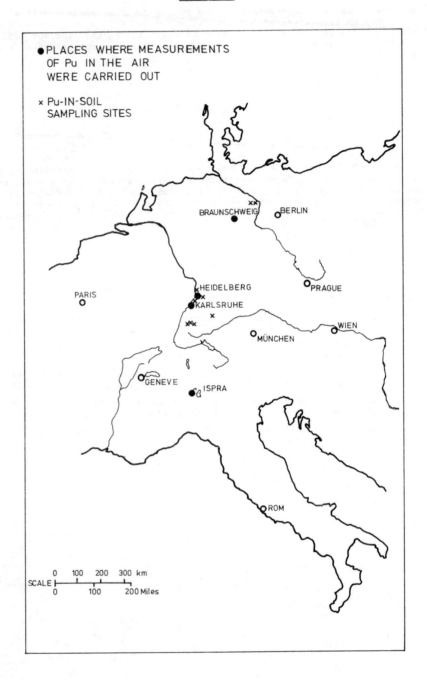

Figure 2

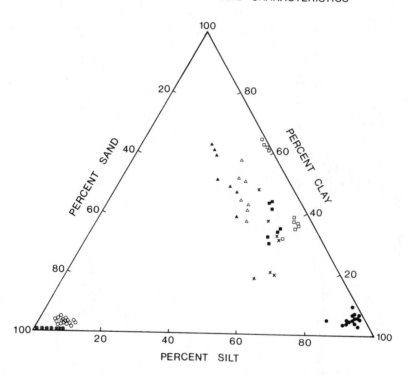

Pu–SAMPLING SITES SOIL CHARACTERISTICS

MINERALS	LOCATIONS
○ QUARTZ	GOLF LINKS
● QUARTZ + ILLIT	GAUANGELLOCH MEADOW
□ ILLIT + CHLORIT	MUSTHOF FARM
▲ MONTMORILLONIT	HOHENBOL HILL
△ ILLIT + MONT. + CHLORIT	BOEHRINGEN QUARRY
■ KAOL.+ MIXED LAYER	SCHOEMBERG ORCHARD
× ILLIT	SCHRAMBERG MEADOW
● QUARTZ	KRAUTZE MEADOW

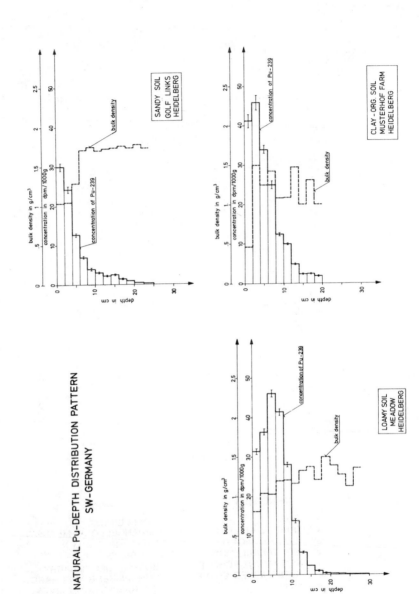

Figure 3

NATURAL Pu-DEPTH DISTRIBUTION PATTERN
SW-GERMANY

Figure 4

NATURAL Pu-DEPTH DISTRIBUTION PATTERN

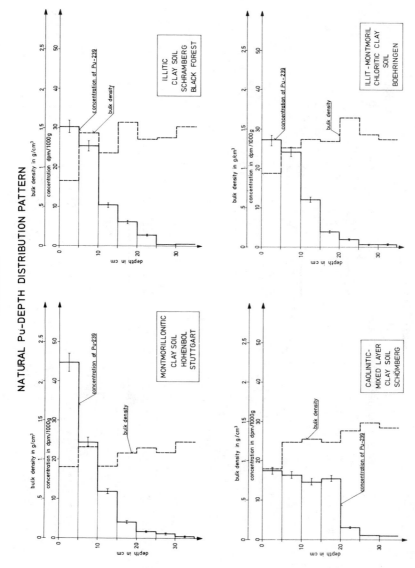

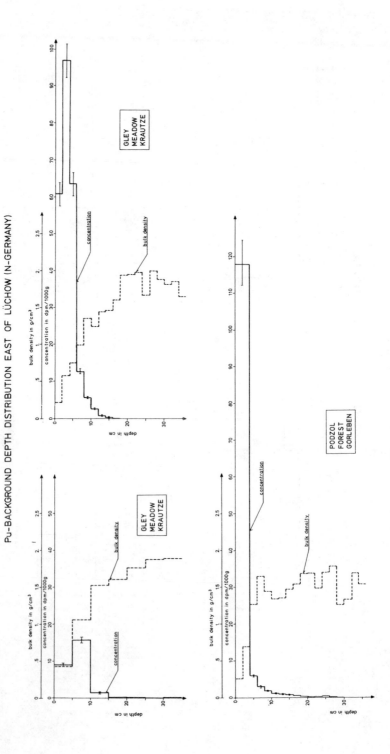

Figure 5

Pu-BACKGROUND DEPTH DISTRIBUTION EAST OF LÜCHOW (N-GERMANY)

Figure 6

HISTORY OF FALLOUT PLUTONIUM MIGRATION IN THE VICINITY OF GORLEBEN (N-GERMANY, NEAR BRAUNSCHWEIG)

TIME IN YEARS →

DEPTH IN CM ↓

Depth (cm) \ Year	1954	1955	1956	1957	1958	1959	1960	1961	1962	1963	1964
2.00	0.0253	0.0498	0.0851	0.1178	0.1756	0.2292	0.2024	0.1806	0.1698	0.4008	0.5097
3.75	0.0	0.0042	0.0121	0.0251	0.0421	0.0670	0.0982	0.1215	0.1387	0.1522	0.2029
6.28	0.0	0.0	0.0005	0.0017	0.0044	0.0087	0.0154	0.0252	0.0370	0.0499	0.0636
8.67	0.0	0.0	0.0	0.0000	0.0001	0.0003	0.0008	0.0015	0.0027	0.0044	0.0066
9.88	0.0	0.0	0.0	0.0	0.0000	0.0000	0.0001	0.0001	0.0003	0.0006	0.0010
12.13	0.0	0.0	0.0	0.0	0.0	0.0000	0.0000	0.0000	0.0001	0.0001	0.0003
13.62	0.0	0.0	0.0	0.0	0.0	0.0	0.0000	0.0000	0.0000	0.0000	0.0000
15.54	0.0	0.0	0.0	0.0	0.0	0.0	0.0	0.0000	0.0000	0.0000	0.0000
17.88	0.0	0.0	0.0	0.0	0.0	0.0	0.0	0.0	0.0000	0.0000	0.0000
19.88	0.0	0.0	0.0	0.0	0.0	0.0	0.0	0.0	0.0	0.0000	0.0000
21.96	0.0	0.0	0.0	0.0	0.0	0.0	0.0	0.0	0.0	0.0	0.0000
23.51	0.0	0.0	0.0	0.0	0.0	0.0	0.0	0.0	0.0	0.0	0.0
26.18	0.0	0.0	0.0	0.0	0.0	0.0	0.0	0.0	0.0	0.0	0.0
27.77	0.0	0.0	0.0	0.0	0.0	0.0	0.0	0.0	0.0	0.0	0.0
	0.0253	0.0540	0.0977	0.1446	0.2222	0.3053	0.3169	0.3290	0.3486	0.6080	0.7841

Depth (cm) \ Year	1965	1966	1967	1968	1969	1970	1971	1972	1973	1974	1975	1976
2.00	0.5206	0.4580	0.3919	0.3512	0.3049	0.2669	0.2319	0.1973	0.1680	0.1486	0.1275	0.1090
3.75	0.2665	0.3249	0.3666	0.3927	0.4091	0.4160	0.4158	0.4097	0.3985	0.3835	0.3670	0.3487
6.28	0.0820	0.1063	0.1356	0.1677	0.2009	0.2340	0.2660	0.2963	0.3242	0.3493	0.3714	0.3906
8.67	0.0093	0.0126	0.0168	0.0221	0.0285	0.0359	0.0442	0.0532	0.0626	0.0724	0.0823	0.0921
9.88	0.0015	0.0023	0.0033	0.0045	0.0061	0.0080	0.0103	0.0130	0.0161	0.0195	0.0232	0.0270
12.13	0.0005	0.0009	0.0014	0.0022	0.0032	0.0045	0.0061	0.0082	0.0107	0.0138	0.0174	0.0214
13.62	0.0001	0.0001	0.0002	0.0003	0.0005	0.0008	0.0011	0.0016	0.0021	0.0028	0.0037	0.0047
15.54	0.0000	0.0000	0.0001	0.0001	0.0002	0.0004	0.0005	0.0008	0.0011	0.0016	0.0021	0.0028
17.88	0.0000	0.0000	0.0000	0.0000	0.0001	0.0001	0.0002	0.0003	0.0004	0.0006	0.0009	0.0012
19.88	0.0000	0.0000	0.0000	0.0000	0.0000	0.0001	0.0001	0.0002	0.0002	0.0004	0.0005	0.0007
21.96	0.0000	0.0000	0.0000	0.0000	0.0000	0.0000	0.0001	0.0001	0.0002	0.0003	0.0004	0.0006
23.51	0.0000	0.0000	0.0000	0.0000	0.0000	0.0000	0.0001	0.0001	0.0002	0.0003	0.0004	0.0007
26.18	0.0	0.0000	0.0000	0.0000	0.0000	0.0000	0.0000	0.0000	0.0001	0.0002	0.0003	0.0004
27.77	0.0	0.0	0.0000	0.0000	0.0000	0.0000	0.0000	0.0000	0.0000	0.0000	0.0001	0.0001
	0.8805	0.9052	0.9159	0.9410	0.9536	0.9667	0.9764	0.9807	0.9845	0.9932	0.9971	0.9999

THE VALUES OF Pu-AMOUNT ARE NORMALIZED, $M_{TOTAL} = \Sigma M(1978)_N = 4{,}875 \times 10^7 \frac{Bq}{km^2}$

Figure 7

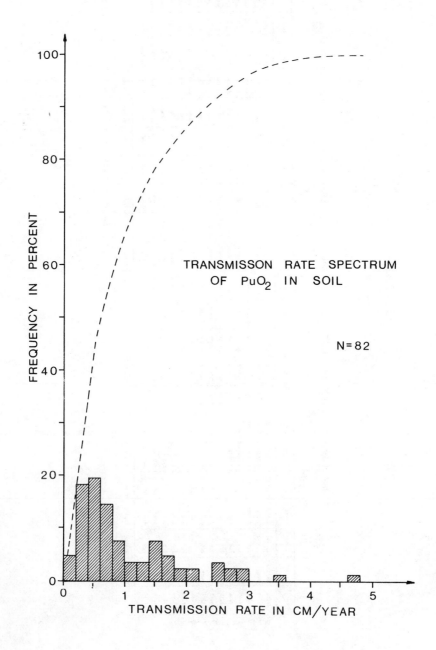

TRANSMISSON RATE SPECTRUM
OF PuO_2 IN SOIL

N=82

Figure 8

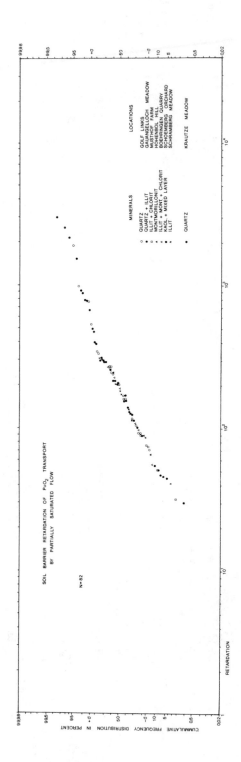

SOIL BARRIER RETARDATION OF PuO_2 TRANSPORT
BY PARTIALLY SATURATED FLOW

N=82

MINERALS

QUARTZ
QUARTZ + ILLIT
ILLIT + CHLORIT
MONTMORILLONIT
ILLIT + MONT. + CHLORIT
KAOL. + MIXED LAYER
ILLIT

QUARTZ

LOCATIONS

GOLF LINKS
GAUANGELLOCH MEADOW
MUSTHOF FARM
HOHENBOL HILL
BOEHRINGEN QUARRY
SCHOEMBERG ORCHARD
SCHRAMBERG MEADOW

KRAUTZE MEADOW

CUMMULATIVE FREQUENCY DISTRIBUTION IN PERCENT

RETARDATION

Figure 9

PREDICTION OF THE ACCUMULATION, TRANSFER AND DISPERSION OF A LOCAL Pu-ANOMALY AT THE GORLEBEN SITE

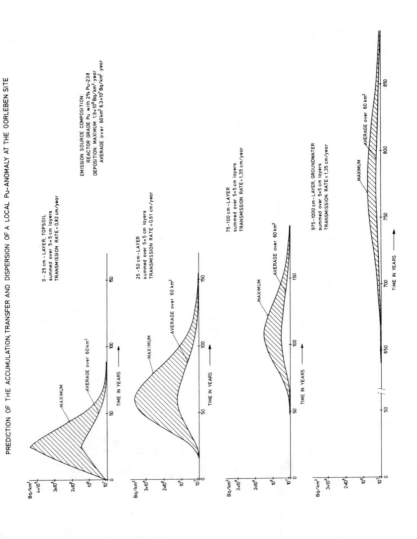

Discussion

G. MATTHESS, Federal Republic of Germany

     Did you take into account in your model the pH and Eh
conditions in the different soil layers ? We have learned that
oxidation and pH control the behaviour, thus these data are of great
importance for the model.

A.T. JAKUBICK, Federal Republic of Germany

     I agree with your opinion. Most likely data on Eh would
improve the correlation, however it is questionable if this improvement is so vital to justify these measurements.

G. MATTHESS, Federal Republic of Germany

     Since Eh and pH are controlling parameters, I suggest that
they should be measured.

P.P. de REGGE, Belgium

     The retardation factor, in the range 50-2000, reported by
the author is in agreement with results, in the low concentration
range, reported in our paper for plutonium solutions and clay samples
taken at - 204 m, showing Kd values in the range 700 to 3000.

G. UZZAN, France

     L'orateur pourrait-il nous préciser si dans le cadre des
mesures de plutonium dans différentes couches du sol, il a également
mesuré le plutonium des espèces végétales poussant sur ces échantillons de sol ? Il serait en effet intéressant de comparer les facteurs
de transfert trouvés dans ce cas avec les facteurs de transfert que
l'on trouve habituellement dans la littérature.

A.T. JAKUBICK, Federal Republic of Germany

     No, not in this case.

D. RAI, United States

     What hypotheses do you make to explain the movement of
plutonium in soils ? How do the soil properties influence the movement ?

A.T. JAKUBICK, Federal Republic of Germany

     The advantage of the black box model approach is that you
do not need any hypotheses. The results suggest however that the
interaction is of electrostatic nature.

M. ALBERTSEN, Federal Republic of Germany

     One point of your work is to calculate the migration
velocities in deeper part from measurements only in the upper soil.
Did you consider that the migration velocities at depth are much
higher than in the A-horizon of the soil with its higher carbon
content ?

A.T. JAKUBICK, Federal Republic of Germany

In the calculation presented I did not consider the differences caused by the organic carbon content. Some more recent calculations I made do consider this effect too. Preliminary results indicate that above a concentration of about 3 % of organic carbon, the effect of organic matter is overwhelming.

P. PEAUDECERF, France

Je voudrais demander au Dr. Jakubick si les mesures de concentration du plutonium ont été réalisées à la même époque de l'année : en période de recharge en eau ou d'évapotranspiration. En effet, les écoulements verticaux saisonniers peuvent modifier grandement la répartition des produits dans le sol.

A.T. JAKUBICK, Federal Republic of Germany

The date of sampling would play a role, I agree. Still, I hope, this effect is negligible, since we are using values averaged over almost 30 years.

J. ROCHON, France

Le facteur retard que vous introduisez dans votre modèle des mélangeurs en cascade est purement empirique et ne repose sur aucun fondement chimique. Pensez-vous que si le cultivateur, que vous avez montré, arrosait son champ de nitrate, vous pourriez prévoir de la même façon la migration du plutonium ?

A.T. JAKUBICK, Federal Republic of Germany

It would very probably influence the migration. To what extent, I do not know. I did not account for it in the model. Nevertheless the results seem to be homogeneous, as the plot of the retardation factors show.

F.P. SARGENT, Canada

I have a comment. You have essentially fitted some observations to a model but taken no account of chemical effects. Also the removal of plutonium by vegetation referred to previously has been ignored.

It seems to me you may well have more than one species migrating and what you see is the more soluble form that has gone further, while the less soluble one has stayed behind and it is purely coincidence that you have a confidence factor of .95.

A.T. JAKUBICK, Federal Republic of Germany

Yes, I did forget to refer to this point. The thing is that the correlation equation I obtained has a threshold value. Now this threshold value can change and go down to 1 just like you would expect. This is where your different species might fit ; but the bulk of the activity moves as described so the amount of the other species is quite small and we do not need to consider it.

<u>D. RAI</u>, United States

Did you measure the concentration of plutonium in equilibrating solutions of contaminated soils ?

<u>A.T. JAKUBICK</u>, Federal Republic of Germany

No. Mainly because of the very low level of plutonium concentrations. We are dealing with concentrations in the order of 10^{-14} molar. However, we do sorption studies in the lab with the same material.

LABORATORY INVESTIGATIONS ON THE RETENTION OF FISSION PRODUCTS BY GRANITES*)

M.Tschurlovits
Atominstitut der Österreichischen Universitäten
A-1020 Vienna, Austria

Abstract

The retention properties for radionuclides in ground water of some Austrian granites has been investigated with a simple static laboratory technique, the elements being Sr, Cs, and Ce, so far. The measured K_d range from 30 to 9000 for Ce, from <2 to 15 for Sr, and from 60 to >2500 for Cs. Different granites show different retention properties, the influence of other parameters such as pH being smaller in relation.

*)This work was suggested by the UNSCEAR Secretariat

1. Introduction

The assessment of human exposures due to the migration of radionuclides from a granite respository of high level wastes requires
information on the retardation of their transport respect to ground
water. As water movement through granite is mainly by fissure flow,
surface phenomena are important in determining apparent distribution
constants for different radionuclides.
The retention properties can be studied either in situ, comparing
the arrival time of different nuclides, or with a sample of a well
defined grain size, and from such samples, the retention as a function of area can be derived.
Results obtained by static laboratory methods for some Austrian
granites will be presented in this paper. The elements investigated
have been Ce, Sr, and Cs, as a just stage of the project, which plans
to cover also actinides and their daugthers.
The reason to apply the second (static) method in the reported work
is that simpler equipment and lower labelling activity are necessary
for the determinations.

2. Experimental Procedures

Granite samples from different sites (Table 1) were ground, obtaining
a broad spectrum of grain sizes. To obtain well defined grain size
samples, test sieves[1] with a width of mesh between 144 and 324
[cm^{-2}] was used for the present experiments.
Samples of 10 grams of each granite powder with the grain size mentioned above were put into 250 ml plastic bottles, together with
100 ml of distilled water. The radioactive solution with suitable
activity concentration was then added, and the bottle closed and
shaken. The activity was chosen to obtain suitable statistical precision (ranging from a few to 100 µCi per bottle) and from each
granite, 3 samples were prepared to reduce pipetting and other
errors to a minimum. The samples were shaken in regular intervals.
In addition, to determine an eventual interaction of the activity
with the plastic bottle, 3 blank samples without granite were prepared in each series.
After the time allotted for interaction and retention, an amount of
100 µl of the water from the granite and the blank samples was taken,
put on a counting planchets and dried. The planchets were counted
with a NaJ(Tl)-detector and MCA, and the results were corrected for
background and compton contributions, respectively.
Due to the application of Ce-144 (134 keV), Sr-85 (514 keV), and
Cs-137 (660 keV), no problems with interference of these radionuclides in the NaJ(Tl)-detector occurred.
The concentration factor K_d, which is described in the annex, was
obtained easily by determination of the ratio of the water activity
concentration at the granite with the activity concentration of the
blank, and no problems with different measuring sample properties
occurred. For the conditions of these experiments the concentration
factor K_d in granite is obtained by

$$K_d = (\frac{a_b}{a_s} - 1).10 \tag{1}$$

where a_b ... activity concentration in blank sample

a_s ... activity concentration in water at granite

[1]DIN 1171

3. Results

3.1 Sites of Granite Samples

Samples were taken from the sites shown in Table 1 without specific selection.

sample code	site	country	remarks	color
1	Grein, Aumühle	OÜ.	fresh broken,	light grey
2	Herrschenberg Waldviertel	NÜ.	available as cutted, stone slab (with 3 cm thickness)	light grey
3	Neustadel Platte	NÜ.	fresh broken,	
4	Gebharts, Waldviertel	NÜ.		dark grey
5	Neuhaus	OÜ.		light grey
6	Kastenhof	OÜ.	surface, weathered	light brown
7	Grein, Aumühle	OÜ.		brown
8	Gloxwald	OÜ.		dirty grey brown

NÜ ... Lower Austria
OÜ ... Upper Austria

3.2 Strontium

Table 2 shows the results of the measurements for a_b/a_s with Sr as $SrCl_2$ (pH $\sim$ 4), expressed as ratio of activity concentration blank sample to activity concentration water from granite sample.

sample code ↓→	time [h]			
	4	56	150	340
1	1,22	1,20	1,29	1,4
2	1,25	1,24	1,32	1,4
3	1,31	1,28	1,49	1,5
4	1,20	1,29	1,31	1,4
5	1,25	1,23	1,43	1,5
6	1,10	1,02	1,03	1,1
7	1,56	1,65	2,22	2,5
8	1,28	1,09	1,13	1,3

Table 2: Activity ratio blank/water of granite for the different samples. Sample code see Table 1.

As easy to see from the table, no significant differences were observed, except for samples 6 and 7, and a concentration factor K_d of about 4 ± 1 can be derived for the granites 1 - 5 and 8.

3.3 Cesium

Tables 3.1 and 3.2 show the results of the ratio a_b/a_s for Cs-137 (CsCl), the pH-values being 4 for table 3.1 and 3 for table 3.2, respectively.

sample code	time [h] 4	56	150	340
1	3,8	14	> 50[2]	> 50
2	2,4	5	10	10
3	2,0	16	> 50	> 50
4	2,1	22	> 50	> 50
5	1,8	6	> 50	> 50
6	1,9	5	20	15
7	2,5	> 50	> 50	> 50
8	2,2	9	36	33

Table 3.1: Ratio a_b/a_s for the samples

sample code	time 160h	284h	310h
1	32	26	40
2	10	14	13
3	55	107	90
4	15	57	125
5	26	34	> 250
6	5,6	5,8	5,5
7	100	210	> 250
8	24	45	60

Table 3.2: Results for a_b/a_s of measurements with higher activity concentration

As an average over granites 1, 3, 4, 8, a concentration factor K_d of about 750 ± 350 can be derived. The samples 5 and 7 show signifi-cant better retention properties, the concentration factor K_d being larger than 2500. Sample 2, but in particular sample 6, have lower concentration factors K_d of 130 and 60, respectively.

3.4 Cerium

Table 4 shows the results of the ratio a_b/a_s obtained for Ce-144 (CeCl$_3$ in HCl, pH being $\sim$3).

[2] This value was obtained by estimation of the deviation of the back-ground in the energy region considered. It was estimated for 2 x 95% confidence interval, being equal to 4.65 $\sqrt{B}$, where B background events in the counting time.

sample code	160h	284h	310h
1	61	44	70
2	35	56	67
3	91	157	220
4	23	88	350
5	68	200	200
6	3	3	3
7	420	670	$\sim$900
8	33	38	50

Table 4: Ratio a_b/a_s for Ce versus time of interaction

For this element, no significant groups can be distinguished except
the by far best properties of granite 7 having an equilibrium concentration factor K_d of about 9000, and K_d of sample 6 is much smaller
as for the other samples. The remaining samples show nonuniform
properties. Averaging over the results for samples 1, 2, 3, 4, 5,
and 8 a concentration factor of approximately 1600 ± 1000 can be
derived.
In another series (pH 4) all samples except 6 show concentration
factors K_d of greater than 200, sample 6 a factor of 30.

4. Discussion and Conclusions

	Ce	Sr	Cs
1	700	4	400
2	650	4	130
3	2200	5	900
4	3500	4	1250
5	2000	5	> 2500
6	30	< 2	60
7	9000	15	> 2500
8	500	3	600

Table 5: Results for equilibrium concentration factor K_d

Table 5 presents the results of the investigations so far. The
granites show, except for the poor results of the weathered   sample 6
and the sample 7, which has considerable better properties, concentration factors ranging from 500 to 3500 for Ce, 3 to 5 for Sr and
130 to 1250 for Cs. If the last results are grouped to similar
values, groups of $500/650/700 \rightarrow 620 \pm 100$; and $2000/2200/3500 \rightarrow$
2500 ± 800 for Ce, and $400/600/900 \rightarrow 600 \pm 250$; and larger than 2500
for Cs appears.
Sample 7, which is a light brown granite with feldspar, quartz and
mica particles, has the by far best properties of the samples
investigated within this paper, and the better properties can be
recognized easily by faster retention of the radionuclides.
Sample 6, a weathered granite from the surface, has much poorer
retention properties, probably due to the environmental influences.
From the   results obtained in this work, the influence of the pH-
value seems to be smaller as the differences in the granites their-
self. The results are in the same order of magnitude as Swedish
results [1].

For the future it is planned to extend the measurements for other parameters and for other radionuclides, which are contributing significantly to a collective dose commitment as well as an improvement of the method, if necessary.

Reference

[1] Kärnbränslecykelns slutsteg, Förglasat avfall fran upparbetning, IV Säkerhetsanalys, Sweden, 1977

The author is very indebted to Dr.Dan Beninson, UNSCEAR Secretary, for suggesting this work and for helpful discussion, and also to Mr.W.Rupp for his assistance in performing the measurements.

Appendix

Theory [1]

The retardation of radionuclides being in ground water by granite is described by a factor (retardation factor), which is defined as:

$$K_i = \frac{\text{velocity of water}}{\text{velocity of radionuclide}}$$

and that is for a given movement equal to

$$K_i = \frac{\text{time of radionuclide migration}}{\text{time of water migration}}$$

The retardation factor can be written as:

$$K_i = 1 + K_a \cdot a_1$$

where K_a: equilibrium constant $[\frac{Ci}{m^2} \text{ granite}/\frac{Ci}{m^3} \text{ solution}]$,

a_1: area available for ion exchange $[m^2 \text{ granite}/m^3 \text{ solution}]$

K_a can be obtained by

$$K_a = \frac{K_d}{a_2} \quad [\frac{Ci/kg \text{ granite}}{Ci/m^3 \text{ solution}} / \frac{m^2 \text{ granite}}{kg \text{ granite}}]$$

where K_d is the equilibrium concentration factor, determined in the experiment and a_2 being the area per unit mass specific for the experiment $[m^2 \text{ granite}/kg \text{ granite}]$.

Discussion

P. BO, Denmark

Do you know the concentration dependence of the Kd values of your granite samples ?

M. TSCHURLOVITS, Austria

The concentration dependence is not yet determined. The concentration in the experiments was given by the carrier free activity necessary for counting (2-100 µCi/bottle).

A.A. BONNE, Belgium

Les résultats indiquent l'importance de la préparation des échantillons pour de telles expériences. Regardons l'échantillon 6, qui est un granite altéré. En broyant et tamisant la roche, on perd une fraction importante pour la sorption, notamment les argiles d'altération, peut-être aussi des minéraux comme l'oxyde de fer. Les résultats de mesures du Kd pour l'échantillon 6 sont très bas et ceci peut être dû à la perte de la fraction argileuse.

M. TSCHURLOVITS, Austria

It may be possible that the method of sample preparation is responsible for the poor Kd of sample 6. But, in an early experiment with sample 2, the dependence of Kd from the grain size was assessed, and in equilibrium no differences appeared, but the retention happened faster for the small grain size.

RADIONUCLIDE TRANSPORT THROUGH HETEROGENEOUS MEDIA

J. Hadermann
Swiss Federal Institute for Reactor Research
CH-5303 Würenlingen (Switzerland)

ABSTRACT

A semi-analytic solution for radionuclide transport from a nuclear waste repository through layered geological media with piece-wise constant parameters is presented. The properties of the solutions are discussed. Numerical results are given and discussed for several realistic sets of relevant parameters.

1. INTRODUCTION

Radioactive wastes generated by the nuclear power economy are planned to be disposed into geological formations. Generally a leach incident is considered [1] as the critical event in a risk analysis for a waste repository. Consequently radionuclide transport through the geological formation to the biosphere is of crucial importance.

Solving the radionuclide transport equation, two approaches have been pursued. First, numerical solutions in various dimensions (see e.g. ref.[2]) and, second, analytical solutions in homogeneous media (see e.g. ref.[3]). In the present paper(1) a solution of the one-dimensional radionuclide transport equation in heterogeneous geological media is presented. The different layers are assumed to be characterized by piece-wise constant parameters. As a consequence a semi-analytic solution can be written down. Furthermore, we choose the initial and boundary conditions such that in the first layer a fully analytical solution results. This allows for a detailed discussion of the resulting expressions. The advantages are prevailing for several reasons. First, the main characteristics of the transport problem and the boundary conditions at the repository are taken into account. Second, the influence of different parameters can be investigated more easily. Third, a parameter study can be performed with reasonable expenditure. The last two points are of special importance since the parameters entering the transport equation are subject to extreme uncertainties.

2. SOLUTION OF THE TRANSPORT EQUATION

Under the assumption of a linear sorption isotherm the radionuclide transport equation is given [4] by

$$R(\lambda C + \frac{\partial C}{\partial t}) = \vec{\nabla} \cdot (D\vec{\nabla}C) - \vec{\nabla} \cdot (\vec{v}C) \qquad (2.1)$$

where C is the radionuclide concentration, λ the radionuclide decay constant and $\vec{v}$ the water velocity. R is the retention factor [5] which may vary over decades, thus giving rise to the main uncertainty for the solution of equation (2.1)

For not too small water velocities the components of the dispersion tensor are given [6] by

$$D_{ij} = a_T v \delta_{ij} + (a_L - a_T) \frac{v_i v_j}{v} \qquad (i,j = x,y,z) \qquad (2.2)$$

where a_T and a_L are transversal and longitudinal dispersivities, respectively.

In full generality, eq.(2.1) requires numerical solution making parameter studies extremely expensive. Studies of ground water flow show that longitudinal convection and dispersion are generally much greater than transverse. For this reason the one dimensional restriction of eq.(2.1) is considered to be a good approximation provided the migration path is long compared to the waste repository's dimensions. Consequently, introducing the nuclide velocity $u=v/R$ and writing $a=a_L$ eq.(2.1) reduces to

---

(1) A more detailed version will appear elsewhere.

$$\frac{\partial C}{\partial t} = au \frac{\partial^2 C}{\partial z^2} - u \frac{\partial C}{\partial z} - \lambda C \qquad (2.3)$$

where z is the coordinate of the linear migration path. In accordance with the scope of the present work we divide the migration path into N parts (2) of lengths ℓ_i with piece-wise constant parameters a_i and u_i defining new coordinates z_i (see figure 1). In each section i the concentration $C_i(z_i,t)$ is a solution of equation (2.3).

The initial and boundary conditions are taken to be

$$C_i(z_i,0) = 0 \qquad (2.4)$$

$$C_i(\infty,t) = 0 \qquad 1 \le i \le N \qquad (2.5)$$

and (3) $\quad C_i(0,t) = C_{i-1}(\ell_{i-1},t). \quad 2 \le i \le N \qquad (2.6)$

What remains is the boundary condition at the repository. We will consider (4) two separate expressions:

$$C_1(0,t) = C_0(1-s(t-T))e^{-\lambda t} \qquad (2.7a)$$

$$C_1(0,t) = C_0' t(1-s(t-T))e^{-\lambda t} \quad , \qquad (2.7b)$$

where s is the unit step function and T the leach time. The rationale for the ansatz (2.7) is twofold. First, it allows for an analytic solution in the first layer and thus assures comparison with well-known analytic homogeneous models (e.g. ref.[8]). Relinquishing this condition, one may use any boundary condition at the repository. Second, the main characteristics of nuclide concentration at the repository are included, namely radioactive decay, finite leach time and, by combination of (2.7a) and (2.7b) reduction of waste inventory through leaching.

Defining the function $G_i(z_i,t)$ by

$$C_i(z_i,t) = G_i(z_i,t)e^{-\lambda t} \qquad (2.8)$$

it can be shown that the solution of eqs.(2.3)-(2.7) is given by the Laplace transformed function

$$\mathcal{L}[G_i(z_i,t)] = \mathcal{L}[G_1(0,t)]e^{\alpha_i z_i} \prod_{k=1}^{i-1} e^{\alpha_k \ell_k} \qquad (2.9)$$

---

(2) One may think of different geological layers or of a single layer with piece-wise different physico-chemical properties such as porosity or sorption.

(3) From the method of solution it becomes evident that without any difficulty this could be changed to

$$C_i(0,t) = \tilde{\gamma}_i C_{i-1}(\ell_{i-1},t)$$

with $\tilde{\gamma}_i$ constant. Thus, irreversible loss of radionuclides at layer boundaries could be taken into account.

(4) For the special case $C_1(0,t) = C_0$ and $\lambda=0$ a solution of eqs. (2.3)-(2.6) has been given by Shamir and Harleman (ref.[7]) in the context of mixing of miscible fluids.

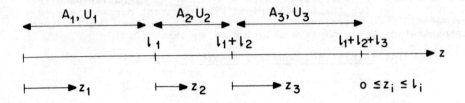

Figure 1: Definition of coordinates for a migration path in piece-wise homogeneous media.

TABLE I : Parameters characterizing the different layers. It is recalled that only the combination $v/(1+K_d\rho_b/\varepsilon)$ enters the transport equation.

Layer	Porosity ε	Distribution Coeff. [mℓ/g] K_d (Sr)	Distribution Coeff. [mℓ/g] K_d (Cs)	Water Velocity [m/d]	Nuclide Velocity
Clay	0.01		1000	0.01	L
	0.05		500	0.02	M
	0.1	10	100	0.1	H
Granite	0.05		1000	0.1	L
	0.1		100	0.3	M
	0.2	1	10	1.0	H
Sandstone	0.1		500	0.1	L
	0.15		100	0.2	M
	0.3	1	50	1.0	H

where $\mathcal{L}$ denotes the operator of Laplace transformation and α_j is given by

$$\alpha_j = \frac{1}{2a_j} \left(1 - \sqrt{1 + \frac{4a_j}{u_j} p} \right) \qquad (2.10)$$

Consequently, with the aid of the convolution theorem of Laplace transformation, the concentration in layer i can be calculated recursively

$$C_i(z_i,t) = \int_0^t C_{i-1}(\ell_{i-1},\tau)K_i(t-\tau)d\tau \qquad (2.11)$$

where the kernel is given by

$$K_i(t) = \frac{z_i}{2\sqrt{\pi u_i a_i t^3}} \exp\left\{ -\lambda t - \frac{(u_i t - z_i)^2}{4u_i a_i t} \right\} . \qquad (2.12)$$

Two important properties for the solution (2.11) follow from the general structure of eq.(2.9).

First, the solution is invariant with respect to permutation of layers. This property can be of use in a numerical integration in eq.(2.11). Second, since α_k is independent of ℓ_k, layers with the same parameters a_k and u_k can be combined to a single layer of corresponding length. In effect, a complicated "sandwich structure" may be combined, e.g., into a simple three-layer-problem, thus reducing the numerical expenditure appreciably.

3.   DISCUSSION OF THE SOLUTIONS

In this paragraph we will discuss some properties of the solutions given in the last section. In order to be specific and not obscure the main points by too lengthy expressions we will restrict ourselves to two layers.

Writing in detailed notation

$$C(\ell_2,t,T,u_2,a_2): = C_2(\ell_2,t)$$

we remark that the concentration, in principle, depends on the nine parameters $\ell_i, u_i, a_i, t, T, \lambda$. However, except for trivial factors, the five dimensionless parameters

$$\beta = \frac{a_1}{u_1} \frac{u_2}{a_2} \qquad \gamma_i = \ell_i/a_i \qquad \delta = \frac{u_2}{a_2} t \qquad \Delta = \frac{u_2}{a_2} T , \qquad (3.1)$$

are determining the concentration. Introducing these parameters we have

$$C(\ell_2,t,T,u_2,a_2) = \frac{\gamma_2 \beta}{2\sqrt{\pi}} e^{-\lambda t} \int_0^{\delta/\beta} D_1(x) \frac{\exp\left\{ -\frac{(\delta-\beta x-\gamma_2)^2}{4(\delta-\beta x)} \right\}}{\sqrt{(\delta-\beta x)^3}} dx$$

$$(3.2)$$

with

$$D_1(x) = \frac{C_o}{2}\left\{ \text{erfc}\left(\frac{\gamma_1-x}{2\sqrt{x}}\right) + e^{\gamma_1}\,\text{erfc}\left(\frac{\gamma_1+x}{2\sqrt{x}}\right)\right.$$

$$\left. -s(\delta-\Delta)\left[\text{erfc}\left(\frac{\gamma_1-(x-\frac{\Delta}{\beta})}{2\sqrt{(x-\Delta/\beta)}}\right) + e^{\gamma_1}\,\text{erfc}\left(\frac{\gamma_1+(x-\Delta/\beta)}{2\sqrt{x-\Delta/\beta}}\right)\right]\right\}$$

(3.3a)

for the boundary condition (2.7a) and

$$D_1(x) = \frac{C_o'}{2}\frac{\beta}{\delta}\,t\left\{(x-\gamma_1)\,\text{erfc}\left(\frac{\gamma_1-x}{2\sqrt{x}}\right) + (x+\gamma_1)e^{\gamma_1}\,\text{erfc}\left(\frac{\gamma_1+x}{2\sqrt{x}}\right)\right.$$

$$-s(\delta-\Delta)\left[(x-\gamma_1)\,\text{erfc}\left(\frac{\gamma_1-(x-\Delta/\beta)}{2\sqrt{x-\Delta/\beta}}\right)\right.$$

$$\left.\left. +(x+\gamma_1)\,e^{\gamma_1}\,\text{erfc}\left(\frac{\gamma_1+(x-\Delta/\beta)}{2\sqrt{x-\Delta/\beta}}\right)\right]\right\}$$

(3.3b)

for the boundary condition (2.7b). The integral in eq.(3.2) is invariant under certain transformations of the original parameters. This gives rise to scaling laws exhibiting the dependence on the parameters.

3.1 UNDER THE TRANSFORMATION

$$u_i' = ku_i, \qquad t' = t/k, \qquad T' = T/k \tag{3.4}$$

where k is an arbitrary constant, we have

$$C(\ell_2, t/k, T/k, ku_2, a_2) = \exp\left\{-\lambda(1/k-1)t\right\}C(\ell_2, t, T, u_2, a_2) \tag{3.5a}$$

$$C(\ell_2, t/k, T/k, ku_2, a_2) = \frac{1}{k}\exp\left\{-\lambda(1/k-1)t\right\}C(\ell_2, t, T, u_2, a_2) \tag{3.5b}$$

This clearly demonstrates the importance of small nuclide velocities introducing essentially an effective decay constant λ/k. On the other hand one sees that an uncertainty in the nuclide velocity is equivalent to an uncertainty in the time scale.

3.2. UNDER THE TRANSFORMATION

$$a_i' = ka_i, \qquad t' = kt, \qquad T' = kT, \qquad \ell_i' = k\ell_i, \tag{3.6}$$

we have

$$C(k\ell_2, kt, kT, u_2, ka_2) = \exp\left\{-\lambda(k-1)t\right\}C(\ell_2, t, T, u_2, a_2) \tag{3.7a}$$

$$C(k\ell_2, kt, kT, u_2, ka_2) = k\exp\left\{-\lambda(k-1)t\right\}C(\ell_2, t, T, u_2, a_2). \tag{3.7b}$$

Thus, an uncertainty in the dispersivities is reflected in an uncertainty in the time scale as well as the migration lengths in the layers. As a result, for fixed migration lengths ℓ_i, small values of the dispersivities may not be conservative assumptions.

3.3 THE TRANSFORMATION

$$a_2' = ka_2, \qquad u_2' = ku_2, \qquad \ell_2' = k\ell_2 \tag{3.8}$$

yields the result

$$C(k\ell_2, t, T, ku_2, ka_2) = C(\ell_2, t, T, u_2, a_2) . \tag{3.9}$$

A question of particular interest is the possibility of homogenization. This can be attained only [5] if $\beta=1$, i.e. the very special relation

$$\frac{a_1}{u_1} = \frac{a_2}{u_2} \tag{3.10}$$

holds. Under this condition the transformation (3.8) with $a_2' = a_1$ and $u_2' = u_1$ leads to

$$C(\frac{u_1}{u_2} \ell_2, t, T, u_1, a_1) = C(\ell_2 t, T, u_2, a_2) \tag{3.11}$$

where the left hand side clearly is the solution of the homogeneous problem with a total migration length $\ell = \ell_1 + (u_1/u_2)\ell_2$.

4. NUMERICAL RESULTS AND CONCLUSIONS

Numerical calculations have been done for several nuclides. We show some results for ^{135}Cs, a nuclide with appreciable sorption. Three different geological media are considered: (i) clay (or a layer with strong clay content), (ii) granite and (iii) sandstone. For each layer water velocity, effective porosity and distribution coefficients were varied within reasonable limits such that low (L), medium (M) and high (H) radionuclide velocities result (see table I). As a standard configuration (a) we assumed: migration path lengths ℓ_1 = 500 m in clay, ℓ_2 = 2000 m in granite and ℓ_3 = 8000 m in sandstone; dispersivity a_i = 30 m and leach time T = 10^3 a. Starting from these values the following variations were made: (b) ℓ_2 = 200 m, ℓ_3 = 800 m; (c) a_i = 3 m; (d) T = 10^5 a. All other parameters are fixed to the values of the standard configuration for the variations (b)-(d).

The boundary condition at the repository was taken to be the sum of equations (2.7a) and (2.7b) with C_0' = -10^{-5} C_0 (time taken in units of years), giving rise to a linear decrease of the nuclide concentration at the repository. Thus, in cases (a)-(c) a small part of the nuclide inventory is leached out, only, whereas in case (d) the whole inventory is leached out.

Figures 2 - 6 show results of the present calculations. Except for figures 3 and 5 the ratio $C_3(\ell_3, t)/C_0$ is displayed as a function of time for the different variations (a)-(d) and selected

---

(5) Of course, in a general sense homogenization is always possible. However, here the point is not to derive relations which are equally or even more laborious to calculate than the original expression (3.2).

- 239 -

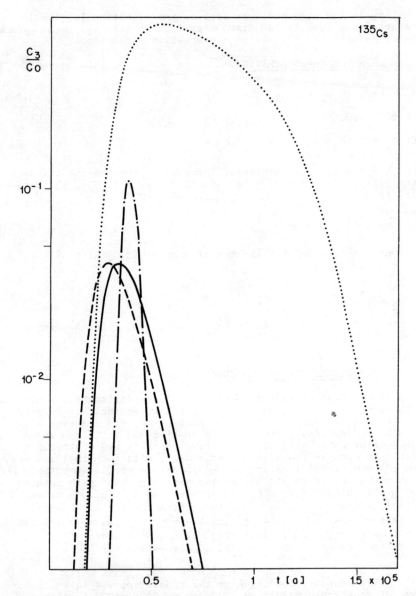

Figure 2: The ratio $C_3(\ell_3,t)/C_0$ as a function of time.
Curves are for different parameter sets: (a) full line,
(b) dashed line, (c) dash-dotted line and (d) dotted line
as described in the text. The nuclide velocities are H,H
and H, respectively, for the three layers (see table I).

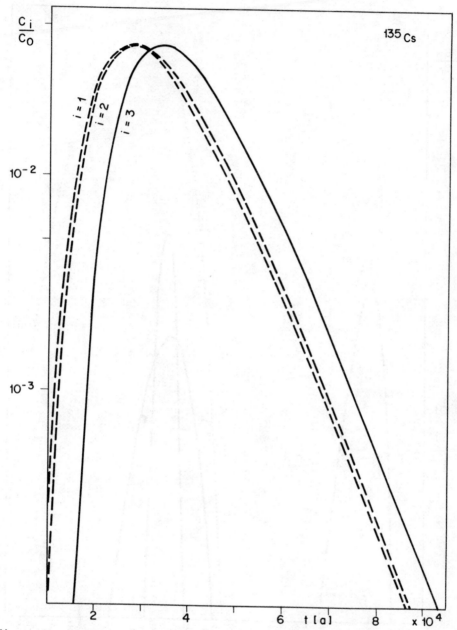

Figure 3:   Effect of the three layers. The dashed lines are
             $C_1(\ell_1,t)/C_0$ and $C_2(\ell_2,t)/C_0$, respectively, whereas the
             fully drawn line corresponds to the full line in figure 2.

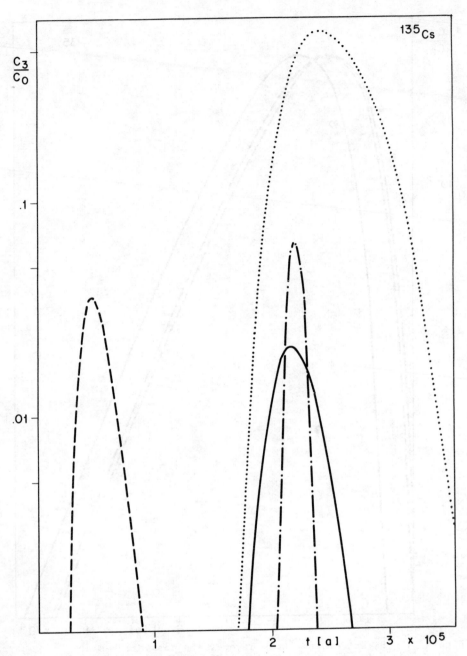

Figure 4: Same as figure 2, but for nuclide velocities H,M and M in the three layers.

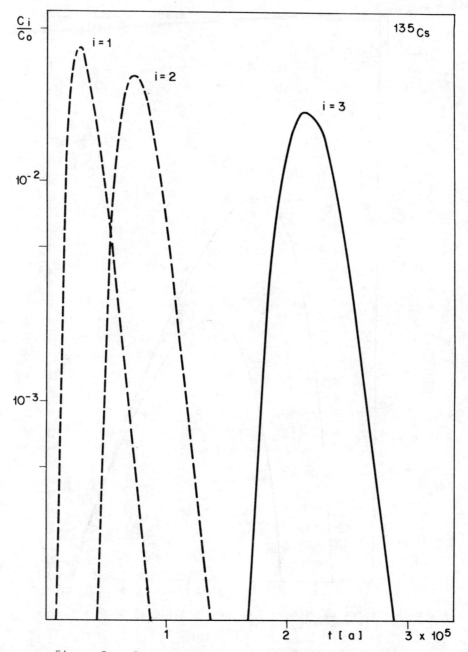

Figure 5: Same as figure 3, but for set (a) in figure 4.

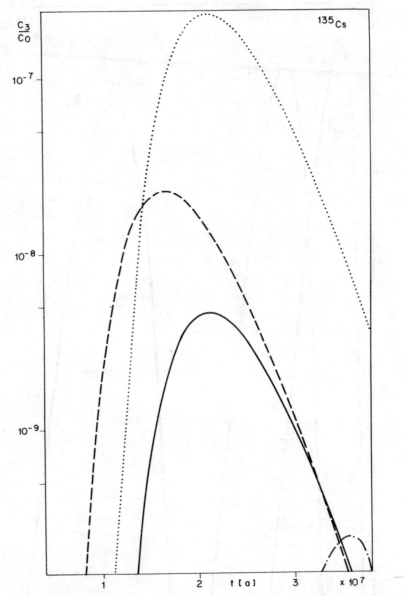

Figure 6:   Same as figure 2, but for nuclide velocities L,L and L
            in the three layers.

values of nuclide velocities (L,M,H) in the three layers. This ratio can be considered as an attenuation factor generated by the combined effect of retardation, dispersion and radioactive decay. In figures 3 and 5 the effect of the different layers is exhibited.

Depending on the parameters, the ratio $C_3(\ell_3,t)/C_0$ varies over seven decades. The following conclusions can be drawn from these results:

(i) If one of the layers (in our case clay) shows much smaller nuclide velocities than the others, this layer dominates to the extent that migration in the other layers can essentially be neglected. This can be seen in figure 3.

(ii) If the nuclide velocities are not too different, each layer contributes to the attenuation depending on the migration length, figure 5.

(iii) If migration time is much larger than half-life, also layers with relatively high nuclide velocities may contribute (fig. 6) and small dispersivities are no longer conservative estimates.

These conclusions, (i)-(iii), are valid not only for [135]Cs but quite generally. They show that for case (i) a homogeneous calculation taking into account the most important layer (6), only, would suffice. In contrast cases (ii) and (iii) require a heterogeneous calculation. Though, at a first glance it may seem that a homogeneous calculation could reproduce the results, it has to be pointed out that a detailed reproduction of the heterogeneous calculations is difficult and even in an approximate fit the corresponding parameters have no simple relation to the original ones. Therefore, in these cases a heterogeneous calculation is much more reliable and efficient.

Concluding we want to mention two restrictions of the present approach. First, transverse dispersion is not included. In this respect the calculations yield conservative results. Second, decay chains are not included. These may become important for actinide migration. Further investigation on these lines is in progress.

ACKNOWLEDGEMENT

The author would like to thank J. Patry for helpful discussions.

REFERENCES

[1] Report to the Am.Phys.Soc. by the study group on nuclear fuel cycles and waste management, Rev. Mod.Phys. 50 (1978) S1.

[2] Schwartz, F.W.: J. Hydrol. 32 (1977) 257.

[3] Baetsle, L.H.: Nuclear Safety 8 (1967) 576.

[4] See for example Borg, I.Y. et al.:"Information Pertinent to the Migration of Radionuclides in Ground Water at the Nevada Test

(6) We recall that in the present approach a layer may be composed of several spacially separated parts.

Site", UCRL-52078, Part 1, University of California, Lawrence Livermore Laboratory (May 1976).

[5] Giddings, J.C.: "Dynamics of Chromatography", Dekker New York 1975.

[6] Scheidegger, A.E.: J. Geophys. Res. 66 (1961) 10.

[7] Shamir, U.Y. and Harleman, D.R.F.: J. Hydraulics Division, Proc. Am.Soc. Civil Eng. 93 (1967) 237.

[8] Burkholder, H.C. et al.: "Incentives for Partitioning High-Level Waste", BNWL-1927, Battelle Pacific Northwest Laboratories (Nov. 1975).

Discussion

P. BO, Denmark

Do your scaling factors have a physical meaning or are they only a mathematical property of the equations to be solved ?

J. HADERMANN, Swtizerland

The scaling laws are a consequence of the fact that the solution is dependent on the five parameters (3.1) instead of the nine-where λ gives a trivial dependence-original ones. The three scaling laws involve different sets of physical parameters.

E. BÜTOW, Federal Republic of Germany

It seems that, for your numerical solution method, a stepwise input data sensitivity analysis could be performed to find the range of the possible critical hydrological, chemical or physical radionuclide transport conditions. Have you found a critical parameter constellation ?

J. HADERMANN, Swtizerland

The time needed for a single calculation is very small, around 10 seconds on a CDC 6500. This allows for a large variation of the parameters taken into account. In general, a critical parameters constellation cannot be given since the importance of a parameter is dependent on nuclide, migration path, and the parameters of the geological layers considered.

A. SALTELLI, CEC

I think that your treatment of scaling laws is interesting, but in my opinion analytical solutions are not so easy to handle as you say ; in particular if the radionuclide under consideration is not the first term of a decay chain. I think that for this kind of calculations numerical solutions are better.

J. HADERMANN, Switzerland

As stated in the paper the present model does not include decay chains. Essentially it is appropriate for fission products only. I think we need both : analytical and numerical models. Analytical models are suited for analyzing certain parameters since they allow for very time efficient parameter studies.

INVESTIGATIONS ON THE MIGRATION AND THE RELEASE OF WATER WITHIN ROCK SALT

N. Jockwer
Gesellschaft für Strahlen- und Umweltforschung mbH München
Institut für Tieflagerung - Wissenschaftliche Abteilung
Berliner Str. 2, 3392 Clausthal-Zellerfeld / Germany

Zusammenfassung

Steinsalzformationen, in denen hochradioaktive Abfälle eingelagert werden sollen, enthalten zwischen 0,1 und 1,0 Gewichtsprozent Wasser. Dieses Wasser wird bei Temperaturen über 100 ^{o}C freigesetzt. Hinsichtlich der langfristigen Sicherheit und der Ausbreitung der Radionuklide ist dieses Wasser von Bedeutung, da es zu chemischen und physikalischen Wechselwirkungen zwischen den Abfallbehältern und dem Verfestigungsprodukt kommen kann und es Transportmedium für die Radionuklide ist.

Abstract

Rock-salt formations, which shall be used for disposing of high-level radioactive waste, can contain between 0,1 and 1,0 weight-percent of water. This water will be set free at temperatures above 100 ^{o}C and may migrate on the boundary surfaces, the micro-fissures and the intergranular spaces. In regard to the long term safety and the migration of longlived radionuclides, this water is important, because it involves chemical and physical interactions with the waste-containers and the solidified waste-products and it may be a transport-medium for the radionuclides.

1. The different forms of the water within the rock-salt

The water content of a great number of samples of Older and Younger Halite with different mineralogical compositions have been determined. This investigation showed that the water within the rock-salt exists in four forms:

1.1. Hydration-water of the various minerals

This water component may be very important to the high-level waste disposed, because its amount is proportional to the mineral components. The release temperature of this water form depends on the mineral constituent

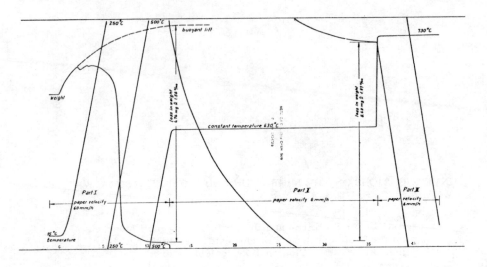

Figure 1: Thermogravimetric plotter diagram of rock-salt

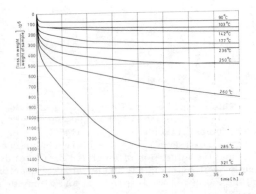

Figure 2: Loss in weight at various constant temperatures
as a function of time

(i. e. carnallite, polyhalit, gypsum, forwhich it is released between 70 and 200 °C). After being released, the water may migrate on the boundary surfaces, the microfissures and the intergranular spaces with high velocity.

1. 2. The "crystal boundary surface-water"

This water form has a temperature of release less than 200 °C and a high diffusion velocity on the boundary surfaces, throngh the microfissures and the intergranular spaces. It may migrate to the heat sources by thermo-diffusion and a vapour pressure gradient.

1. 3. Brine dorplets within negativ crystals (inclusions)

In rock-salt from the Asse II Salt Mine, these inclusions mainly exist in formations were solution metamorphism took place. Generally, the diameter of these inclusions are between 0, 01 to 0, 1 mm, some may be up to 5 mm.

1. 4. Water molecules trapped within the crystal lattice

This water contributes very little to the total water content. These water molecules are fixed within the crystal lattice and their diffusion velocity is very low. Therefore, their temperature of release is comparantively high. Consequently this water form is least important to the disposed high level waste.

2. Thermodynamic effects

This section will deal mainly with the termogravimetric investigation of the rock-salt. This method shows what happens with the water, when the temperature increases.

For the thermogravimetric determination of the water content with a thermoanalyser, about three grams of ground rock-salt is placed in a ceramic pot, is heated at a constant time-temperature rate, and the sample weight is plotted continuously (figure 1). In this diagram, three different weight-loss-temperature-zones are shown. Different heating rates were chosen to show different effects.

Temp. Zone I: The heating velocity was 5 °C/min. The apparent increase in weight is caused by the temperature because the density of the air decreases with increasing temperature and the buoyant lift becomes less. The loss in weight above 100 °C is caused by the release of water.

Temp. Zone II: At 630 °C the temperature has been halted constant for 40 hours. This further loss in weight is caused by a water and gas release (i. e. H_2S, CO_2).

Temp. Zone III: After 40 hours when the weight became constant the temperature was raised to 730 °C. This loss in weight is caused by sublimation.

In order to determine the influence of the temperature to the salt and water content, the loss in weight as a function of time at differet constant temperatures between 90 and 350 °C has been measured. Some of these curves are shown in figure 2. The loss in weight is normalized by sample weight, in order to compare different salttypes. This diagram shows that at temperatures less than 300 °C the weight did not become constant within 40 hours. It can not be said that after a very long time at temperatures between 80 and 250 °C, the whole water would be set free or whether there will be different steps. This figure also shows that for nearly the same temperatur-intervals, the distants between the curves are not constant. When the loss in weight after 35 hours is plotted as a function of temperature (in figure 3), it shows that increasing temperature frees the water within the rock-salt in steps. It could not be said that a relationship exists between these steps and the mineral components. Further investigations are necessary with salt samples of different mineral components.

In order to investigate the diffusion and migration of the released water in the rock-salt, cylinders with 15 cm of length and 7 cm of diameter have been prepared. Brass tubes were shrunken around the cylinders, so that release of water is possible only through the ends of the samples.

These samples were put into an oven at different temperatures for various times. After drying, they were cut into disks and their water contents were analysed. A concentration gradient has not been measured within the cylindrical samples. However, the final water content was much lower than befor drying. After 6 hours of drying at a 280 °C, they lost about 46 %; after 12 hours, about 64 % of the original water. In the thermogravimetric analyser (see figure 2), a ground sample at 285 °C lost about 52 % within 6 hours and 72 % within 12 hours. These results indicate that the water escaping the crystal-system, can diffuse along the crystal boundaries and the microfissures.

As the released water migrates on the boundary surfaces and the microfissures, thermodiffusion of this water may be the mechanism. Therefore, we will expose rock-salt samples to a temperature gradient and see if the water migrates to one end. These investigation will start in 1979.

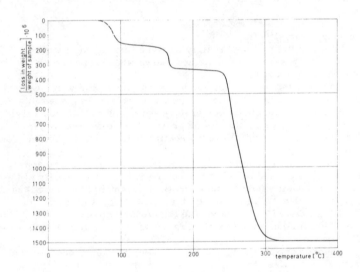

Figure 3:   Loss in weight after 35 hours as a function
             as a function of temperature

Discussion

J. HAMSTRA, The Netherlands

Brine migration and cristal water release at elevated rock salt temperatures are subjects of interest because they may lead to restrictions for the thermal loading of a high-level waste burial area. The investigations described were performed on rock salt samples at atmospheric pressure and mostly at temperatures beyond the boiling point of water. There was no temperature gradient in the sample. The driving mechanism for the release of water must thus have been the pressure difference between the water vapour formed and the outside atmosphere.

A comparable condition exists in a burial area only immediately after disposal of the high-level waste canisters in a borehole. Shortly, borehole convergence will restore a lithostatic pressure in the rock salt surrounding the buried canisters. At elevated confining pressure there will be no pressure difference, only a temperature gradient in the rock salt.

Therefore the investigations described are, to my opinion, not representative for the conditions that exist in a rock salt burial area.

Do you agree with my comment ?

Do you plan to perform in-situ brine migration tests at a later stage, or, more generally, do you think that such in-situ tests are feasible ?

N. JOCKWER, Federal Republic of Germany

I think that these investigations are representative for a salt burial area. Around a borehole, in which high-level waste has been disposed of, a temperature gradient will exist and therefore a vapour pressure gradient will exist too. This vapour pressure gradient leads to migration. To study this effects in the laboratory we expose samples of rock salt to a temperature gradient and observe water migration.

Furthermore, I think that in the disposal area and around the borehole the lithostatic pressure does not exist at the beginning. For this period of time the laboratory tests will show the right behaviour of the released water.

Last year we made an in situ experiment in which the released water was measured. Two electric heaters of 2 kW each were placed in a borehole about 5 m deep. The wall of the borehole was heated up to 130°C. Within 120 days we found that about 40 ml of water had been released into the borehole.

Further in situ experiments on water migration will be carried out.

L.R. DOLE, United States

I would like to make some comments. I guess there was a reference to the duration of the thermal pulse. Depending on high-level waste or spent fuel being present in the repository, the pulse may last 200 to 600 years, which in a geological sense is a short time. Our modelling of brine migration based on laboratory measurements of the movement of brine inclusions covers the first 50 years of the thermal exposure of the face of the salt adjacent to the waste. As far as re-establishing lithostatic pressure at the bore-hole face adjacent to the canister our programme in the US has dealt with a 20-25 year retrievable phase during which there is essentially atmospheric pressure in the borehole. So these pressure differentials in case of retrievable geologic disposal can last for quite some time ; essentially for the most sensitive period from the point of view of the influx of brine. Therefore brine migration might be a problem in regard to corrosion rates of the waste canisters and retrievability of the waste. We do not know for sure what the brine migration rate is ; we have some calculations. Our experience as well as some of the experience of the Asse mine show that at least the initial rates of influx of brine are somewhat lower than theo-retically predicted. We presume this is due to the compressive stresses around the heated holes. Certainly in Project Salt Vault, where the atmosphere in the hole was monitored, the most significant influx of brine took place right after the heaters were turned off ; in other words when the heat flux was relaxed, there was an opening of the interfaces of the salt crystals and the largest influx of moisture. I think in-situ tests are ultimately required because of the complex relationship between the stresses around the borehole, the heating rates and the thermal fluxes involved.

V.E. Della LOGGIA, CEC

I would like to ask you if during the in situ experiment that you have mentioned, it was possible to see any effect of the water, as far as migration of radionuclides is concerned.

N. JOCKWER, Federal Republic of Germany

We have not yet studied the migration of radionuclides in situ.

GENERAL DISCUSSION DISCUSSION GÉNÉRALE

F. GIRARDI, CEC

The speakers of this session have raised issues which have partially been touched upon previously. The first speaker and the following discussion have addressed the question of the usefulness of studying existing phenomena, such as fallout, in order to understand mechanisms and processes controlling the migration of radionuclides in the environment. We should certainly take advantage of these data; perhaps we should integrate these studies with some laboratory experiments to validate the conclusions. The second speaker has addressed the important question of experimental methodology. The advantage of having standard procedures may again be raised here as a follow-up of this second paper. The third paper has described a method to assess the migration of ions in the geosphere. Two alternative approaches: the analytical and the numerical ones have been discussed. The conclusion that we should have both is certainly satisfactory from a scientific point of view, but programme managers may have practical difficulties with this solution. The last presentation has touched an important issue: if water migrates to the waste canisters, what will be the effect on the disposal site? This again raises the question of how much we should rely on lab tests versus in situ experiments. Of course, it is a matter of optimising available resources.

F. VAN DORP, The Netherlands

I would like to add something about the use of fallout and natural radionuclides data for studying migration to and through the biosphere. You always have difficulties with laboratory experiments, because it is easy to perform a new laboratory experiment to prove that the values found are too low. You can always find some new value for a parameter, which is higher than the one used before. Therefore, when you use these ever higher values in risk analysis, you will get worse results for normal situations. Perhaps, by using the long-term experiments which are performed by nature with fallout and especially with natural radionuclides, you can find some relation also between laboratory data and in situ data. So, you have a better estimate of the real values for these parameters.

G. MATTHESS, Federal Republic of Germany

I would like to make another comment in the same direction, which covers perhaps the same field of politics of science, or politics of money for science. I think we cannot neglect either one of these different tasks. One point that I have to deal with is the measurement of Kd values and the transfer of these values into a predictive model which, in theory, can be used to describe nuclide migration. The scientific background of Kd measurements is very difficult because, in the laboratory, we reduce all the variation in parameters as much as possible. We restrict ourselves to study fewer models, to produce fewer systems with only one mineral and only one solution, and to work with this controlled system. Such systems are very helpful to understand the specific mechanisms but are not very helpful to predict what is going to happen in nature. Therefore, I think that we have to study the very dirty natural systems. For instance, it is very complex to understand what is going on within such a soil system as described by Dr. Jakubick. We may be very sophisticated in model systems and we know a lot about plutonium oxidation states and so on. But in this soil system, we have a lot of varying parameters. If we try to understand what is going on, we have to make many more measurements than in our laboratory studies. From a practical standpoint, it means that we need much more manpower and much more instrumentation to do in situ measurements. We need these data, because we are not certain that we have introduced the right values into our predictive models.

P. BO, Denmark

I agree to some extent with you that the so-called "right values" can only be obtained in situ. However, in situ experiments will only give you values for one specific site. It is very probable that the predominant parameters, which determine what has actually happened, will vary quite a lot from site to site. For example, you have soils with high and low carbonate content, with high and low ion exchange capacity, with and without organic complexing agents, and so on. It means that the in situ experiments at one site are unreliable for predictions at another site. Laboratory studies are more fundamental in nature. Not because they are laboratory experiments, but simply because it is a way in which you can isolate single effects and order them by importance. Then going to specific sites, you can say that a site has such organic materials and such minerals with an ion-exchange capacity in a certain range. Therefore, we would expect that these effects are important here and some others are important there. This laboratory information is used in the evaluation of in situ experiments. In situ experiments are very important, non only because of the need for direct site data, but also because the whole technique of making in situ experiments needs to be developed further. In situ techniques must be developed and laboratory measurements must be correlated in a reasonable degree with the in situ results. But I still believe that the laboratory experiments will have to be the basis.

A.T. JAKUBICK, Federal Republic of Germany

The conclusion is the same that we reached in connection with the modelling effort. We need both lab studies and in situ studies.

F. GIRARDI, CEC

The philosophy, that we have in our studies, is to try to reproduce in the laboratory, as close as possible, the characteristics of a real situation. Then, we try to understand why we have certain Kd values, certain distribution profiles, and so on. Therefore, we discover a series of mechanisms that will allow us to understand why a site has behaved in a specific way. This does not guarantee, of course, that another site will behave in the same way. But at least, we will be able to make an educated guess on how it might behave. You have the site characteristics on one hand and the laws of chemistry and physics on the other hand. That is why we have set up the type of equipment that Avogadro has described in his paper. We tried to simulate the behaviour of a repository, through which water flowed and then passed through geological media. You have seen in the theoretical studies, which have been motivated by these experiments, how we plan to interpret what happens.

I agree that in situ experiments are expensive. The costs of in situ experiments are so high that they should be done after an educated guess has been obtained from the lab. They should be more a type of verification experiment on natural systems upon which we already have some knowledge. This may raise some more discussion, but this is how we now proceed.

D. RAI, United States

I would like to add that last year the staff of Battelle Pacific Northwest Laboratory evaluated close to about 2.000 literature reports, which attempted to study the effects of various soil properties on nuclide behaviour. In all those studies I regret to say that we found that one parameter or the other was indicated as the most important in a particular soil. It turns out that you can list about 20 soil properties and you can always select a case in which that particular property appeared to affect the results. Therefore, if you want to transfer a specific system result to another situation, you cannot.

L.R. DOLE, United States

I think the philosophy of our programme resulted from the requirement to evaluate a very broad spectrum and a very large number of materials throughout North America. We are required to evaluate many alternatives and model many hypothetical cases. So, our approach was to gather as much detailed empirical parametric data as possible on a broad number of "generic samples", representative of North America. Our strategy was to first do a very rigorous empirical parametric study of many materials. Those data would be useful later in the more detailed mechanistic studies.

F.P. SARGENT, Canada

One of the problems related to the study of near surface effects concerns the prediction of future conditions; for example, in Canada we are interested in the disposal of high-level wastes in a hard rock repository. Presumably, any involvement in near surface phenomena is going to be thousands of years hence. Therefore, it seems to me that we would like to get some feeling of how the soils are going to be in several thousand years time. For example, will they have a glacier on top of them? Some of the things we are talking about today seem somewhat academic in that time context. However, if we are talking about a contemporary fuel reprocessing plant or a land burial site, yes, I can see it.

A. VAN DALEN, The Netherlands

I did not hear all that was said in connection with geochemical studies. Many organisations are looking for mineral deposits. Not only the minerals themselves, but also the accompanying elements, in order to have some indication of the genesis of these deposits. As far as I know, there is a lot of information which may be of value. The migration of accompanying elements is also being investigated, perhaps in geological circumstances similar to a repository. This field of scientific research might provide a lot of useful information.

G. MATTHESS, Federal Republic of Germany

I agree with the different comments regarding the needs for basic laboratory studies. But as a hydrogeologist, I have a

bad feeling that we are doing work which has little to do with reality. I think the in situ programmes, running in Sweden and in
Canada, give us an idea of what is really happening at depth, and
could be representative of a 1000-m-deep repository in hard rock
or salt. I have to make in situ measurements, and then set up laboratory tests.

What is the hydrology of a 600-m-thick clay formation?
It is a tremendous mass of an almost impermeable material; therefore, what is the mass transport in such a system? How do we measure the hydraulic gradients driving water in this system? I cannot imagine it. I have to study it in the field.

P. BO, Denmark

The problem is that probably you could not do it in the
field.

L.R. DOLE, United States

I think part of the problem is that we are trying, in the
consequence analysis of accident scenarios, to look at the perturbed
natural material in some hypothetical situation. I am certain that
none of us would put the waste in any place, where, under normal
conditions, there is an expectancy of large hydrological gradients
or transport. It is in the process of analysing the consequences
of hypothetical events that we address some of these unnatural conditions: i.e. transport through clays with hydrological gradients
that are perhaps very improbable.

M. TSCHURLOVITS, Austria

I think that the decision between laboratory and in situ
measurements should be determined by the final goal of the analysis; i.e. the collective dose resulting from a burial site. Such
analysis cannot be more accurate than one or two orders of magnitude, because there are as many as a hundred mostly unknown parameters in the models. A decision on what is the necessary level
of investigation can be based on the sensitivity of the dose model.

H. BRÜHL, Federal Republic of Germany

I think that the grinding of geological specimens for laboratory study has the risk of operating too far from the natural
conditions. In general, it is preferred to work in the laboratory,
with undisturbed materials, if possible.

F.P. SARGENT, Canada

It seems to me that discussions like this can go on and
on. What we are trying to do is to identify the ranges of the
variables. When we select a site or an appropriate method, we can
factor the ranges into it. You might decide to choose a site that
has very good retention at depth and poor retention at the surface
for some peculiar reason. So what we need is some appreciation of
the range of the variables - both at the surface and at depth.
Therefore, the debate as to whether you do experiments in situ or
in the lab is not really relevant. I mean that obviously you have
to do both. The in situ studies are clearly site specific; the
lab investigations, for example the Kd measurements, are very dependent on the material used. We had a good illustration of that
today where the numbers ranged from 15 to 250 for materials that

were described as granite. In this particular case, we now know that you can get Kd values varying from 15 to 250. We did not know that until we did the experiments. So, we need to do lab experiments. It seems to me that we need in situ experiments too.

G. MATTHESS, Federal Republic of Germany

My previous remark was caused by the data that have been presented by Dr. Rai. He compared all the data obtained by different labs for the same materials. I see that the Atomic Energy of Canada Limited has the lowest values. I was at the AECL last year and I asked them about their low values. The answer was that their values agree well with the observed migration in the field. The real movement of strontium and cesium was exactly the same as observed in the lab. And I think it was this argument which persuaded me to follow the Canadian approach.

L.R. DOLE, United States

I would like to comment on the point made by Dr. Sargent, that we need to know the range of values. I think that it is simply not enough. What we need to have, and unfortunately we do not have yet, is a coherent model for the whole system: cost-benefit analysis, thermo-mechanical, nuclide migration, consequence analysis, so that we can optimize the system, making the right decisions on site location, thermal loading, seconday engineered barriers and so on. That is really the goal we are after. This is a very difficult, very complex model. I see that we are very far from having such a large, coherent model. I think we can model parts of the system, some mechanical aspects, some nuclide retardation, etc. Nobody has come near being able to assess the sensitivity of the whole concept with regard to its parts.

F.P. SARGENT, Canada

All that I would like to add to that is that we, as scientists, must appreciate that the final choice will not be made by us. The final choice will be made by politicians. As any of you who has attempted to initiate a drilling programme knows, for example, some communities will have you and some will not. And it is going to be exactly the same for a repository.

L.R. DOLE, United States

But we will be asked to prove unequivocally that the politician's decision was right.

F. GIRARDI, CEC

Are there other comments on the issues raised in this session? I think we have not yet come to the point of giving recommendations, but we are perhaps approaching that moment. I would expect that in the discussion at the end of the last session we might be able to do so. Well if there are no other comments or questions I wish to close this session.

Session V
Ion Exchange and Migration

Chairman — Président
Dr. F.P. SARGENT
(Canada)

Séance V
Échange d'ions et migration

CURRENT STATUS OF NRPB WORK ON THE MATHEMATICAL
MODELLING OF RADIONUCLIDE MIGRATION

Marion D Hill

National Radiological Protection Board, Harwell,
Didcot, Oxon OX11 ORQ, UK

Abstract

The current status of the generic model developed at the National Radiological Protection Board for the prediction of radionuclide migration from geologic disposal sites is described. The model is simple but extremely flexible. It can be used to simulate the migration of radionuclide chains of any length. It has recently been extended to predict migration along flow paths with heterogeneous sorption characteristics. This capability is illustrated by two simple examples. In the first it is assumed that sorption is high close to the geologic repository and decreases along the flow path. This corresponds to the case of a repository which has been back-filled with a material of high sorbent capacity. In the second example it is assumed that there is no sorption of radionuclides in the immediate vicinity of the repository but considerable sorption at greater distances. This provides a means of estimating the potential consequences of large scale fracturing around a repository.

The model can also be used to predict the migration of a radionuclide chain in which one of the members exists in two valence states which are sorbed to different extents. This is illustrated for ^{237}Np and its parent and daughter nuclides. Finally some planned extensions to the model are described.

1. INTRODUCTION

The basic model developed at the National Radiological Protection Board (NRPB) for the prediction of radionuclide migration from geologic disposal sites has been described in a recent report $\underline{/1/}$. It is a simple one-dimensional model which takes into account the major processes occurring during the transport of radionuclides by ground-water, assuming linear equilibrium sorption. It is recognised that this type of model cannot be used to make realistic predictions for a specific geologic site. Its main use lies in generic assessments of the disposal option and in the sensitivity analysis required to indicate areas where further research is necessary and where more realistic modelling will be needed. If a model is to be used in this way it is essential that it should be flexible enough to incorporate all the processes and parameters which may be of interest. For example, a model which is developed on the basis that the rate of release of radionuclides into ground-water is constant could not be used to investigate the sensitivity of calculated doses to the variation of leach rate of high level waste with time after disposal.

The flexibility of the migration model used at NRPB has yet to be fully explored. In particular no attempt has been made to include non-equilibrium sorption, although this could be incorporated into the model, because the data required are not yet available. The model has recently been extended to predict migration along flow paths with heterogenerous sorption characteristics. This capability is illustrated in this paper by two examples. The model can also be used to predict the migration of a chain of radionuclides one of which exists in two different valence states which are sorbed to different extents. This is illustrated for 237Np and its parent and daughter radionuclides. The implications of these results in terms of research and modelling requirements are discussed. The source terms for all the calculations (inventory of radionuclides and rate of release of activity by leaching from vitrified high-level waste) are those used in reference $\underline{/1/}$. The ground-water velocity, dispersion coefficient and length of flow path are also those considered in this report.

2. SPATIAL VARIATION OF SORPTION CONSTANT

Two fairly simple examples will be considered. In the first it is assumed that sorption is high close to the repository and decreases along the flow path. This is the situation which would occur if a repository is back-filled with a material which has a high sorbent capacity for those radionuclides which appear to be poorly sorbed on most geologic media. In the second case it is assumed that there is no sorption in the immediate vicinity of the repository but further along the flow path radionuclides are sorbed to extents typical of a variety of geologic media.

2.1 High sorption in the initial portion of the flow path

For this example it is assumed that the first 600m of the flow path from the repository consists of material on which all radionuclides are strongly sorbed. Calculations are carried out for 99Tc, 129I and the actinide chain of which 237Np is a member. The sorption constants of these radionuclides in the initial part of the flow path are assumed to exceed the values in the remaining part by factors ranging from 10 to 10^5.

The results of the calculations are shown in Tables I and II and illustrated in Figures 1 and 2. The general effects of increased sorption within the first 600m of the flow path are, as expected, decreases in the peak discharge rates of radionuclides at the end of the flow path and increases in the time at which the peaks occur. The peak discharge rates of 99Tc, 129I, 237Np, 245Cm, 241Pu and 241Am decrease by 3-4 orders of magnitude as sorption close to the repository is increased to the maximum value considered. For 233U and 229Th the range of variation in peak discharge rate is small and follows a complex pattern due to the effects of ingrowth from 237Np.

The results show that back-filling a repository with sorbent material decreases the rate of release of radionuclides into ground-water. It also increases the time taken for radionuclides to migrate from the repository to the biosphere. The former effect is of primary importance in reducing the discharge rates of long-lived poorly sorbed radionuclides. In the reference case of uniform sorption along the flow path the migration times of ^{99}Tc, ^{129}I and ^{237}Np are shorter than the time taken for the vitrified waste to dissolve. The discharge rates of these radionuclides are therefore strongly dependent on their rate of release from the waste into ground-water and far less dependent on their migration times. The increase in migration time caused by high sorption close to the repository only has a significant effect on the discharge rates of ^{99}Tc, ^{129}I and ^{237}Np if this time becomes comparable with their radioactive half-lives.

2.2   No sorption in the initial portion of the flow path

In this example it is assumed that no sorption of radionuclides occurs along the first 1 km of the 10 km flow path. The sorption constants along the remainder of the pathlength are taken to be those in the reference cases where uniform sorption is assumed (see Tables II and III. Calculations are carried out for ^{79}Se, ^{93}Zr, ^{135}Cs and the actinide chain of which ^{237}Np is a member.

The results, in terms of peak discharge rates at the end of the flow path are given in Table III. They show that for most of the radionuclides considered the peak discharge rates calculated assuming no sorption at the beginning of the flow path are almost equal to those obtained in the case where sorption is assumed to be uniform along the pathlength. The peak discharge rates of ^{245}Cm, ^{241}Pu and ^{241}Am are approximately a factor of 10 higher than in the reference case. This is due to the small reduction in the migration time of radionuclides. Sorption close to the repository therefore has a very small effect on discharge rates if radionuclides migrate slowly on the remainder of the path to the biosphere.

3.   VARIATION OF SORPTION CONSTANT WITH VALENCE STATES

It has been shown that the sorption properties of some of the actinides vary with valence state $/2,3$ $/$. The radionuclide selected for study in this paper is ^{237}Np. It is assumed that a fraction of the initial inventory of ^{237}Np is in a valence state which is poorly sorbed (sorption constant 100) and the remaining fraction is more strongly sorbed. Sorption constants of 10^3 and 10^4 are considered for this latter fraction. Calculations are carried out assuming the proportions of ^{237}Np in each state vary from 1:9 to 9:1. In each case the migration of the entire actinide decay chain containing ^{237}Np is modelled. Ingrowth of ^{237}Np from its parent radionuclides is assumed to occur into both valence states in the same proportion as they are present in the initial inventory. Sorption constants for the radionuclides other than ^{237}Np are taken to be equal to those in the reference case (see Table II).

The results of the calculations are given in Table IV and illustrated in Figures 3 and 4 for the 1:1 proportion case. They show that the discharge rate of the fraction of ^{237}Np which is poorly sorbed exceeds that of the strongly sorbed fraction over the range of parameters considered. When the sorption constants of the two valence states differ by a factor of 10 the discharge rates of the two fractions will be approximately equal if about 3% of the ^{237}Np is poorly sorbed. When the sorption constants differ by a factor of 100 the discharge rate of the poorly sorbed fraction will be greater than that of the strongly sorbed one if the fraction of the ^{237}Np in the former state exceeds about 0.05%. The results also show how strongly the discharge rates of ^{233}U and ^{229}Th depend on the rates of migration of ^{237}Np. It can be seen from Figure 4 that the discharge rate of ^{233}U increases significantly if a fraction of the ^{237}Np is strongly sorbed.

4.   DISCUSSION

The example calculations described in Section 2 give a broad indication of the effect of sorption within the repository. They show that the sorptive capacity of back-filling and buffer materials for radionuclides which are likely to be poorly sorbed on the geologic media along most of the ground-water flow path must be very high if the peak discharge rates of these radionuclides are to be significantly reduced. However, this conclusion must be viewed with caution because in modelling the migration of radionuclides it has been assumed that sorption is a linear equilibrium process. If irreversible sorption is shown to be the dominant mechanism retardation of radionuclides within the repository could be more significant than indicated by these preliminary calculations.

The results described in Section 3 show the general importance of the variation of sorption constant with the valence state of the radionuclide. They indicate that it will not be sufficient to determine the sorption characteristics of each valence state. It is also necessary to ascertain the fraction of the inventory in each state and how this fraction may change as radionuclides migrate with ground-water from deep to shallow geologic strata. The results also show the influence of the migration rates of parent radionuclides on those of their daughter products. This confirms that it is essential to model the migration of radionuclide chains, rather than individual radionuclides.

5. MODEL DEVELOPMENT

The calculations described in this paper illustrate the capability of the current NRPB model to predict rates of radionculide migration under fairly complex chemical conditions. As indicated in the introduction, this facet of the model will be more fully explored as experimental data on sorption become available and the mechanisms of sorption are elucidated.

The next stage in the development of the model will be to extend it to more dimensions and so simulate radionuclide migration under realistic ground-water flow conditions. However this stage must be approached with caution. Mathematical models are already available to predict ground-water flow patterns around a repository sited in hard rock /4/. These models are capable of including the thermal effects of high-level waste on the permeability of the rock and the influence of thermal gradients on water flow. It is necessary to consider whether radionuclide transport can be included in these models.

There are two main points to be considered. Firstly, the migration of radionuclide chains is extremely complex and it may not be possible to incorporate the detailed chemistry and decay characteristics in a very sophisticated ground-water flow model. This difficulty could be overcome by investigating the sensitivity of the discharge rates of radionuclides to sorption mechanisms and parameters using the type of model used for the present calculations. It may then be possible to make some simplifying assumptions regarding sorption which would enable radionuclide migration to be included in sophisticated flow models.

Secondly, it must be remembered that the reliability of long term predictions of radionuclide migration will always be open to question. It will therefore be necessary to quantify the uncertainties in the analysis and investigate the sensitivity of the results to the assumptions made and the parameters used. The more complex the model the more difficult and expensive it is to carry out a sensitivity analysis. Interpretation of the results also becomes increasingly problematical. It can therefore be argued that simple, robust models have overriding advantages.

These points are raised for discussion purposes; both need to be borne in mind in developing realistic models for generic and site specific assessments.

References

1. Hill, M D and Grimwood, P D. Preliminary Assessment of the Radiological Protection Aspects of Disposal of High-level Waste in Geologic Formations. NRPB-R69. National Radiological Protection Board, Harwell (1978).

2. Fried, S et al. The Migration of Long-lived Radioactive Processing Wastes in Selected Rocks. Argonne National Laboratory ANL-78-46 (1978).

3. Allard, B, Kipatsi, H and Rydberg, J. Sorption av Langlivade Radionuklider i Lera Och Berg. KBS Teknisk Rapport 55. Stockholm (1977).

4. Maini, T and Hocking, G. An Examination of the Feasibility of Hydrologic Isolation of a High Level Waste Repository in Crystalline Rock. Invited paper given at the Annual Meeting of the Geological Society of America, Seattle, Washington (1977).

Table I

Peak discharge rates of ^{99}Tc and ^{129}I calculated assuming higher sorption in the initial portion of the flow path

Radionuclide	Sorption constant in initial portion of flow path	Peak discharge rate (Bq y^{-1})	Time to peak (y)
^{99}Tc	1*	4 x 10^{12}	2 x 10^2
	1 x 10^3	5 x 10^{11}	4 x 10^3
	1 x 10^4	5 x 10^{10}	3 x 10^4
	1 x 10^5	3 x 10^9	2 x 10^5
^{129}I	1*	1 x 10^{10}	2 x 10^2
	1 x 10^3	1 x 10^9	4 x 10^3
	1 x 10^4	1 x 10^8	3 x 10^4
	1 x 10^5	1 x 10^7	3 x 10^5

* Sorption constant assumed for major part of flow path

Table II

Peak discharge rates of ^{245}Cm and its daughter radionuclides calculated assuming higher sorption in the initial portion of the flow path

Radionuclide	Sorption constant in initial portion of flow path	Peak discharge rate (Bq y^{-1})	Time to peak (y)
^{245}Cm	3×10^3 *	8×10	1×10^5
^{241}Pu	1×10^4 *	8	1×10^5
^{241}Am	1×10^4 *	8	1×10^5
^{237}Np	1×10^2 *	8×10^{10}	1×10^4
^{233}U	1.4×10^4 *	9×10^4	2×10^4
^{229}Th	5×10^4 *	6×10^3	4×10^4
^{245}Cm	1×10^4	2×10	2×10^5
^{241}Pu	1×10^4	2	2×10^5
^{241}Am	1×10^4	2	2×10^5
^{237}Np	1×10^4	1×10^9	4×10^4
^{233}U	1.4×10^4	2×10^4	1×10^5
^{229}Th	5×10^4	1×10^3	1×10^5
^{245}Cm	1×10^5	3×10^{-1}	2×10^5
^{241}Pu		3×10^{-2}	2×10^5
^{241}Am		3×10^{-2}	2×10^5
^{237}Np		1×10^8	3×10^5
^{233}U		7×10^5	2×10^6
^{229}Th		6×10^4	2×10^6
^{245}Cm	1×10^6	4×10^{-3}	2×10^5
^{241}Pu		3×10^{-4}	2×10^5
^{241}Am		3×10^{-4}	2×10^5
^{237}Np		7×10^6	2×10^6
^{233}U		6×10^4	3×10^6
^{229}Th		5×10^3	3×10^6

* Sorption constant assumed for major part of flow path

Table III

Peak discharge rates of ^{79}Se, ^{135}Cs, ^{93}Zr and ^{245}Cm and its daughter radionuclides calculated assuming no sorption in the initial portion of the flow path

Radionuclide	Peak discharge rate (Bq y^{-1})	Time to peak (y)
^{79}Se	3×10^{10} (3×10^{10})	9×10^{3} (1×10^{4})
^{135}Cs	2×10^{9} (2×10^{9})	8×10^{4} (9×10^{4})
^{93}Zr	1×10^{9} (1×10^{9})	8×10^{5} (9×10^{5})
^{245}Cm	5×10^{2}	1×10^{5}
^{241}Pu	5×10	1×10^{5}
^{241}Am	5×10	1×10^{5}
^{237}Np	8×10^{10}	9×10^{3}
^{233}U	8×10^{4}	2×10^{4}
^{229}Th	6×10^{3}	4×10^{4}

Notes

Results in parenthesis are those obtained assuming constant sorption along the flow path. The K values used for Se, Cs and Zr were 1×10^{2}, 1×10^{3} and 1×10^{4} respectively. The results for ^{245}Cm and its daughter products calculated assuming uniform sorption are given in Table 2.

Table IV

Peak discharge rates of ^{237}Np and its daughter radionuclides

Fraction of Np in poorly sorbed valence state	Radionuclide	Peak discharge rate (Bq y^{-1})	Time to peak (y)
0.1	^{237}Np (1)	8×10^9	1×10^4
	^{237}Np (2)	3×10^9	9×10^4
	^{233}U	(9×10^3)	(2×10^4)
		3×10^6	1×10^5
	^{229}Th	2×10^5	2×10^5
0.3	^{237}Np (1)	2×10^{10}	1×10^4
	^{237}Np (2)	2×10^9	9×10^4
	^{233}U	(3×10^4)	(2×10^4)
		2×10^6	1×10^5
	^{229}Th	2×10^5	2×10^5
0.5	^{237}Np (1)	4×10^{10}	1×10^4
	^{237}Np (2)	2×10^9	9×10^4
	^{233}U	(4×10^4)	(2×10^4)
		$1 \times 10^6)$	1×10^5
	^{229}Th	1×10^5	2×10^5
0.7	^{237}Np (1)	5×10^{10}	1×10^4
	^{237}Np (2)	9×10^8	9×10^4
	^{233}U	(6×10^4)	(2×10^4)
		9×10^5	1×10^5
	^{229}Th	7×10^4	2×10^5
0.9	^{237}Np (1)	7×10^{10}	1×10^4
	^{237}Np (2)	3×10^8	9×10^4
	^{233}U	(8×10^4)	(2×10^4)
		3×10^5	1×10^5
	^{229}Th	2×10^4	1×10^5

Table IV continued

Fraction of Np in poorly sorbed valence state	Radionuclide	Peak discharge rate ($Bq\ y^{-1}$)	Time to peak (y)
0.5	^{237}Np (1)	4×10^{10}	1×10^{4}
(Sorption constant for strongly sorbed valence state 1×10^4)	^{237}Np (2)	1×10^{8}	9×10^{5}
	^{233}U	(4×10^{4})	(2×10^{4})
		5×10^{7}	1×10^{6}
	^{229}Th	(3×10^{3})	(4×10^{4})
		4×10^{6}	1×10^{6}

Notes

1. Np (1) is the poorly sorbed valence state and Np (2) the strongly sorbed one.

2. The sorption constants assumed for the two valence states of Np are 1×10^{2} and 1×10^{3} unless otherwise stated.

3. The preceding members of the decay chain (^{245}Cm, ^{241}Pu and ^{241}Am) were included in the calculations. Results for these radionuclides are not presented because they are unaffected by variations in the sorption constant of Np.

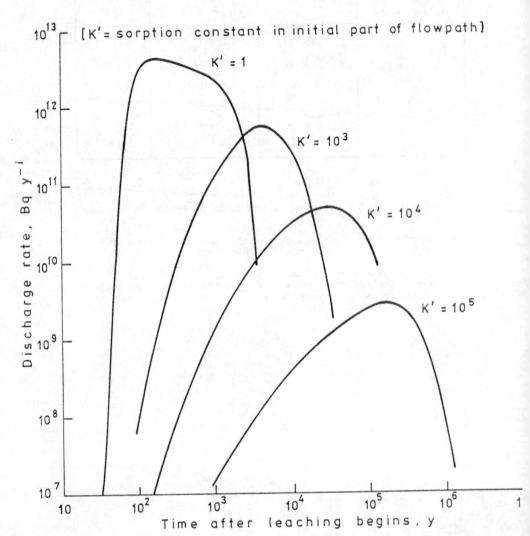

Figure 1. Effect of higher sorption in the initial part of the flowpath on the discharge rate of ^{99}Tc

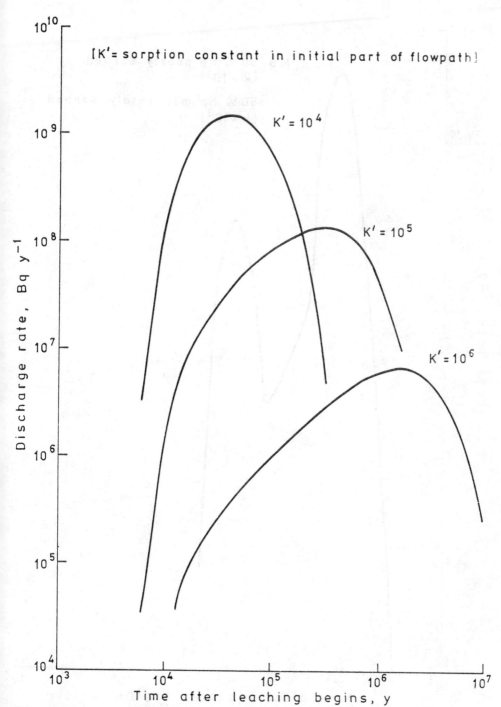

Figure 2   Effect of higher sorption in the initial part of the flowpath on the discharge rate of ^{237}Np

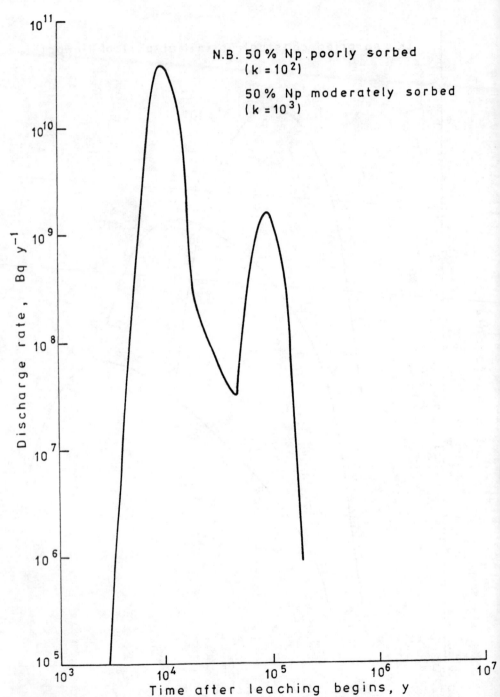

Figure 3. Discharge rates of fractions of ^{237}Np in different valence states

N.B. 50 % Np poorly sorbed ($k = 10^2$)

50 % Np moderately sorbed ($k = 10^3$)

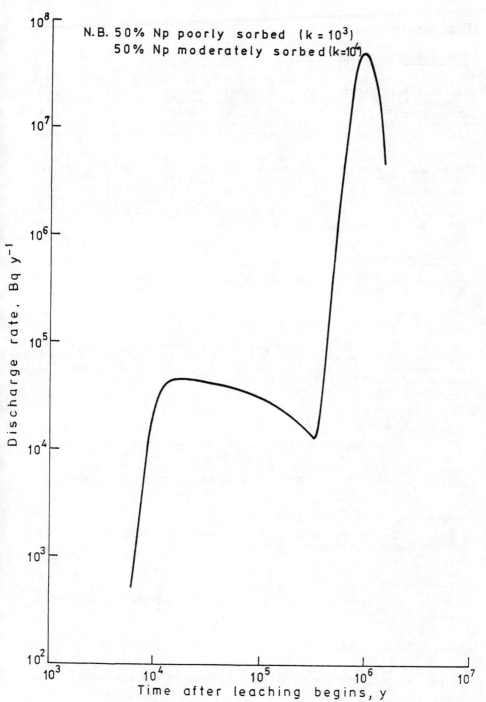

Figure 4. Discharge rate of ^{233}U when ^{237}Np exists in different valence states

Discussion

F.P. SARGENT, Canada

Perhaps, I should mention some of the things that struck me about this paper : First, obviously this is an appeal for the experimentalist to get on with the job and give the modelist some data. Secondly, the use of a model where there is one area with high-sorption followed by an area of low-sorption has important implications for anybody investigating backfill and buffering materials.

Another thing that Miss Hill emphasises is the need that the model should always consider decay chains. At this state of the game we tend to try to make things as simple as possible, however, she has demonstrated that you cannot ignore the chain effects. I like to stress that as a word of warning to people like myself, for example, who tend to take a simplistic point of view and think in terms of just one migrating species.

L.R. DOLE, United States

What was, in your analysis, the source term, or the transfer from waste form to water ?

M.D. HILL, United Kingdom

The waste is assumed to be fractured into pieces of 10 cm diameter and to have a leach rate of 10^{-5} g/cm^{-2}d^{-1}. The source term is therefore a time varying rate of release of radionuclides into ground water. The total time needed by the waste to dissolve is about 3500 years.

L.R. DOLE, United States

Your results confirm H. Burkholder's study : "Incentives for partitioning waste" in that, with long flow paths, the source term relating to waste forms and/or backfill material was not important in relation to concentrations at the point of discharge.

M.D. HILL, United Kingdom

Yes and hence we need to devote more effort to studying nuclide migration than waste forms.

A. SALTELLI, CEC

What kind of grid did you use in performing the integration ?

M.D. HILL, United Kingdom

A fixed grid ; the integration method uses upwind differencing, which gives a very stable solution to the equations.

A. SALTELLI, CEC

In Figure 3 of your paper you show oxidation states of neptunium corresponding to non-equilibrium conditions. Do not you think that in geologic times equilibrium will be attained ?

<u>M.D. HILL</u>, United Kingdom

Equilibrium may well be attained, but we have no evidence to show that this is so and hence we cannot model this situation.

<u>F.P. SARGENT</u>, Canada

Although not relevant to generic models, don't you think that we need to know about statistical failure rate of containers, and its effects on release rates.

<u>M.D. HILL</u>, United Kingdom

The rate of release of radionuclides into ground water has a very small effect on the discharge rates at the end of the flow path if migration times are very long. Statistical failure rate of containers may therefore be of little importance.

<u>F.P. SARGENT</u>, Canada

I have a second question relating to the answer above ; would this apply to technetium and iodine.

<u>M.D. HILL</u>, United Kingdom

No ! Assuming these nuclides are poorly sorbed. The insensitivity of discharge rates to the source term applies only to strongly sorbed nuclides which migrate very slowly.

ION EXCHANGE PROPERTIES OF SOIL FINES

Peter Bo
Risø National Laboratory
4000 Roskilde, Denmark

Right from the first studies of risk analysis in connection with geological disposal of radioactive waste, ion exchange on clay minerals has been considered one of the more important of the phenomena that are responsible for the retention of a number of radionuclides in that part of the geological barrier which consists of argillaceous formations.

In fact, already at an early stage of risk analysis, it was possible, from a few laboratory determinations of distribution coefficients and a general knowledge of ion exchange phenomena, to estimate the order of magnitude and general trends for variation with ground water composition for the distribution coefficient of a number of the chemically more simple radionuclides.

Much of this initial success, I believe, we owe to the general knowledge provided by the very many studies of ion exchange properties of soils made by soil scientists.

It was, however, felt that a fundamental study of the ion exchange properties of clay minerals - with specific relation to the sorption and retention of radionuclides - was needed as part of the basis on which the retention properties of some geological barriers can be determined.

In 1978 we therefore started a programme, under contract with the European Community, with the aim of studying the cation exchange properties of clay minerals in relation to the sorption of radionuclides.

The ultimate goal of this study is to produce a thermodynamic description of the sorption of some typical radionuclides on three different types of clay minerals. The clay minerals chosen will be discussed below. The radionuclides, Cs^+, Sr^{+2} and Eu^{+3} were chosen to represent mono, di- and tri-valent ions, and Cs^+ amd Sr^{+2} are at the same time important components of the radioactive waste.

In order to be useful for the prediction of the sorption of radionuclides under various physico chemical condition, and in order to be able to make a more detailed study of the mechanism of sorption, the thermodynamic description will have to cover a wide range of ground water concentrations and compositions - mainly in terms of the relative amount of mono- and di-valent ions.

We are well aware of that natural ground waters contain many different both mono- and di-valent cations with varying and sometimes very specific affinities for the clay minerals. However, in order to simplify the programme to an extend where it becomes workable without loosing the possibility of studying the general mechanisms of sorption, we decided to work with simulated ground water containing only Na^+ and Ca^{+2} as representatives of the mono- and di-valent cations in natural ground water.

Assuming ion exchange to be one of the mechanisms responsible for the sorption of cations, the Cation Exchange Capacity (CEC) becomes a parameter of prime importance. This is not primarily for sorption-capacity reasons which will at most play a dominant role close to a repository - but more because the CEC, even in that part of the geological barrier where the radionuclides will be present at most in trace amounts, has an important effect on the sorption of these.

As an example, assume that the ground water contains only one cation "k" (Na^+ or Ca^{+2}) with the valence factor "z_k" in a concentration "C_k", and that the clay mineral has a CEC or fixed ion concentration "X". Then the distribution coefficient "K_D^i" of the cationic radionuclide "i" with valence factor "z_i" assumed to be present in trace amounts is given by

$$\log K_D^i = \log \frac{K_i'}{(K_k')^{z_i/z_k}} + (z_i/z_k) \log X - (z_i/z_k) \log (z_k C_k) \quad (1)$$

where $\dfrac{K'_i}{(K'_k)^{z_i/z_k}}$ is the ion exchange selectivity for ion "i"

relative to ion "k". The equation is derived from the simplest possible ion exchange equilibrium and the stipulation of trace amounts of ion "i" is used to simplify the charge balance for the clay mineral, assuming that the number of ions "i" present in the clay is so small compared to the number of ions "k" that they need not be accounted for in the charge balance - i.e. the radionuclide does not occupy any appreciable number of the available ion exchange sites. The above equation shows that, in spite of this, the distribution coefficient is proportional to the cation exchange capacity raised to the power of (z_i/z_k) (x^{z_i/z_k}).

This dependence of the K'_D on the CEC has formed part of the basis on which the three clay mineral types for this study were selected. The clay mineral types chosen were:
a) minerals of the smectite type which according to literature data has a CEC in the range og 80 to 150 meq/100 g clay.
b) minerals of the illite type with a CEC in the range 20 to 50 meq/100 g clay.
c) minerals of the kaolinite type with a CEC in the range 1 to 5 meq/100 g clay.

Other properties which were condensed in the selection of the clay minerals were geometrical arrangement of the ion exchange sites and the possibility of having sites that are very specific for a limited number of cations.

The ion exchange properties of kaolinite is believed to arise from broken bonds and possibly OH-groups on the surface of the particles. This means that the distribution coefficients found on the kaolinite is expected to have a marked dependence on the particle size.

The major part of the ion exchange sites in minerals of the smectite type are due to substitutions of di-valent ions for $Al^{(+3)}$ and di- and tri-valent ions for $Si^{(+4)}$ in the lattice proper - i.e. in the bulk of the mineral. Therefore for this type of mineral the distribution coefficients are not expected to have any appreciable dependence on the particle size as long as the particle size is not very small.

Finally minerals of the illite type are known to have very high selectivity for K^+ and related cations - a phenomenon known as "cation fixation" in the soil sciences.

The programme was initiated by studying the distribution coefficients of Cs^+ and Sr^{++} on the kaolinite in its "pure" Na^+ and Ca^{+2} forms as function of the Na^+ and Ca^{+2} concentrations respectively. The method of preparing these "pure" Na^+ and Ca^{+2} forms of the kaolinite is the standard method in classical ion exchange involving ion exchangers of the "strong acid" type. 0.5 g kaolinite is equilibrated repeatedly with 25 ml 0.1N NaCl or $CaCl_2$ solutions by shaking and the solution is removed by centrifugation and decantation. Following this the samples are equilibrated 4 times with NaCl and $CaCl_2$ solutions of the desired concentration and finally tracer (Sr-85 and Cs-134) were added to the equilibrating solution and the distribution coefficients measured by measuring the activity of the solution and of the kaolinite after 4 hours equilibration and centrifugation to separate the liquid and the solid phase. The reason for measuring directly on the kaolinite, in stead of measuring the decrease in activity of the solution, is that this method gives more accurate results especially when the distribution coefficient and therefore the decrease in solution activity is small. The details of the experiments are described in progress report no. 1 of contract no. 038-78-1 WASDK. The results are shown in fig. 1 through 4, together with the slopes expected from the simplest possible ion exchange description of equation (1) i.e. the factor z_i/z_k in front of the term $\log (z_k C_k)$.

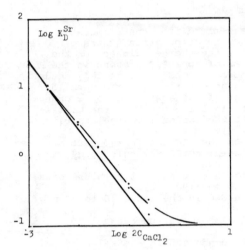

Fig. 1. Distribution coefficient for Sr on Ca-kaolinite.

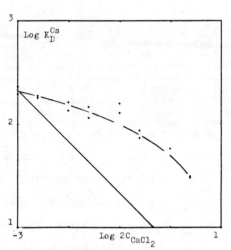

Fig. 2. Distribution coefficient for Cs on Ca-kaolinite.

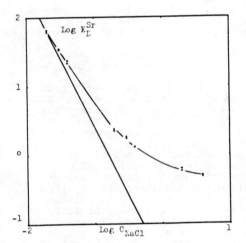

Fig. 3. Distribution coefficient for Sr on Na-kaolinite.

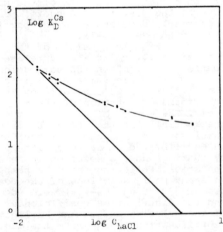

Fig. 4. Distribution coefficient for Cs on Na-kaolinite.

$2C_{CaCl_2}$ means the equivalent concentration of $CaCl_2$.

Starting with the results for K_D^{Sr} it is seen that in neither the $CaCl_2$ nor in the NaCl solutions do the experiments show the "expected" slope but a numerically somewhat lower slope in the low concentration range. This type of behaviour could be explained by assuming that the ion exchange sites on kaolinite are not of the "strong acid" type but of the "weak acid" type and therefore the term

$$(z_i/z_k) \log X$$

in equation (1) is increasing as the concentration of Na^+ and Ca^{+2} are increasing. In fact the addition of NaCl to a slurry of kaolinite makes the pH of the slurry decrease, producing an increased number of dissociated ion exchange sites on the kaolinite. Direct titration of the kaolinite with NaOH and the accompanying increase in distribution coefficients for Cs^+ and Sr^{+2} has also been verified.

The very marked deviations at the higher Na^+ and Ca^{+2} concentrations are probably due to a combination of the above mentioned effect of increasing X with one or both of two additional phenomena. One of these is the concentration dependence of the selectivity ratio

$$\frac{K_i'}{(K_k')^{z_i/z_k}}$$

which contains the activity coefficients of ions "i" and "k" on the clay mineral and in the solution.

The other phenomenon is in classical ion exchange as the breaking down of the "Donnan exclusion" and the invation of salt, as such, in the ion exchanger when the concentration in the solution is increased.

Going on to the Ca-distribution coefficients it is seen from the figures that the deviations from the simple ion exchange behaviour (equation (1)) is even more pronounced than for the Sr-distribution coefficients. But before this behaviour can be explained, an other phenomenon must be examined.

In two experimental determinations of K_D^{Cs} which were believed to be performed under identical physico chemical conditions we found two different values of K_D^{Cs}. This difference we could only trace back to a difference in the Cs^+-carrier concentration used in the two experiments.

Carrier is added in these experiments in order to make sure that the tracers Sr-85 and Cs-134 behave in a thermodynamically predictable way, but the concentrations are kept so low (around 10^{-6} eq/l), compared to the concentration of the "macro" ions Na^+ and Ca^{+2}, that they could not possibly influence the normal ion exchange behaviour of the system. This assumption is the basis for the widely used linear sorption laws in which the distribution coefficient for a given radionuclide is assumed to be independent of the concentration of the radionuclide as long as it occurs only in "trace" amounts.

Having observed the phenomenon we have to explore it further since it will necessarily play an important role in the final thermodynamic description of the sorption of at least Cs^+ on the kaolinite.

The experiments are performed at two concentrations of Na^+ and Ca^{+2} (0.1 and 0.01N) by first converting the kaolinite to the Na^+ and Ca^{+2} form and then equilibrating with the desired Na^+ and Ca^{+2} concentrations as described above. To the final equilibrating solution is added Sr-85 and Cs-134 and varying amounts of Cs- and Sr-carrier and the distribution coefficients are determined as function of these carrier concentrations at the two levels of Na^+ and Ca^{+2} concentrations. The results, reported in progress report no. 1 of contract no. 038-78-1 WASDK, together with the details of the experiments, are shown in Fig. 5 through 8.

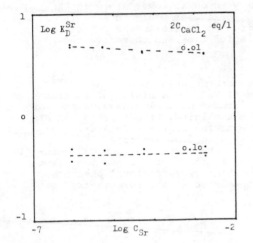

Fig. 5. Effect of Sr-carrier concentration on K_D^{Sr}. Ca-kaolinite.

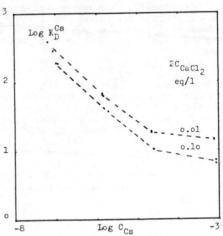

Fig. 6. Effect of Cs-carrier concentration on K_D^{Cs}. Ca-kaolinite.

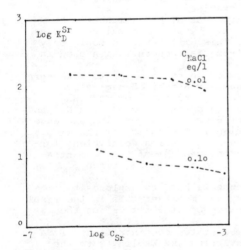

Fig. 7. Effect of Sr-carrier concentration on K_D^{Sr}. Na-kaolinite.

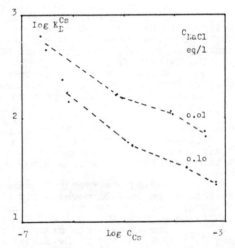

Fig. 8. Effect of Cs-carrier concentration on K_D^{Cs}. Na-kaolinite.

Starting again with the Sr-curves it is seen from the figures that K_D^{Sr} is, especially in $CaCl_2$ solutions, to a good approximation independent of the carrier concentration when this concentration is low compared to the Na^+ and Ca^{+2} concentrations. This means that we have linear sorption laws for tracer amounts of Sr^{+2} on the kaolinite.

In the case of Cs, however, there is a very marked dependence of K_D^{Cs} on the Cs-carrier concentration which levels off at the higher carrier concentrations. Clearly, even trace amounts of Cs^+ does not follow linear sorption laws on the kaolinite.

The mechanism behind this phenomenon is as yet not known, but it may be related to the cation fixation phenomenon known from other clay minerals - mainly of the illite type.

It must also be borne in mind that the clay minerals with which this study is concerned are not completely pure minerals but only minerals that consist predominantly of kaolinite, illite and smectite minerals. A mineralogical analysis of the kaolinite is under way.

Concluding this presentation of the initial experiments and results of the study of ion exchange properties of soil fines we have, in the case of kaolinite seen that, in addition to the parameters which describe classical ion exchange on "strong acid" type ion exchangers, two further parameters are of importance. The first, which was expected on the basis of the structure of the kaolinite, is the pH which together with the Na^+ and Ca^{+2} concentrations determines the degree of dissociation of the fixed ionic groups and thereby the CEC. The pH dependence of ion exchange on "strong acid" type ion exchangers is only important when the H^+-concentration gets so high that normal ion exchange competition with Na^+ and Ca^{+2} ions is appreciable.

The other parameter, which was unexpected in the case of kaolinite, is the Cs-carrier concentration.

Further experiments are needed before the effect of the above mentioned parameters and a thermodynamic description of the sorption of radionuclides on the kaolinite can be given.

In extending the study to other clay minerals we expect that although cation fixation phenomena will occur in minerals of the illite type, the behaviour will, especially for minerals of the smectite type, be much simpler than for kaolinite.

Discussion

D. KINSEY, United Kingdom

It is very good to have unexplained results produced at this meeting. This seems to me to be the proper function of a workshop.

We have preliminary results which indicate that exchange on granite is ion-specific. Exchange of strontium appears to be with calcium only, whereas cesium exchanges with sodium or potassium.

Most of our experiments have been with 3 particle size ranges : < 600 μm, 600 to 2000 μm, 2000 to 3000 μm. There is little difference in sorption of strontium and cesium on particles in these 3 ranges (measured, for example, after 24 hours). However, some tests on powders of ~ 200 μm average particle size showed much higher values of Kd. Going in the other direction, measurements on grains of ~ 20 mm diameter gave very similar Kd values to those for ~ 600 μm powders.

P. BO, Denmark

Your variation in function of particle size agrees with a surface sorption on the particles.

Your observation that strontium exchanges for calcium only and cesium for sodium and potassium only seem very strange from an ion exchange point of view. I would suggest that other processes are responsible for the observed exchanges ; they may be related to surface dissolutions and substitutions.

H.K. BEHRENS, Federal Republic of Germany

The systematic approach of dividing cations into different valence groups seems not to be satisfactory in relation to ion exchange with clay minerals, because of the very high selective reaction with some species, for example cesium. The anomaly of the variation of the Kd of cesium with the cesium concentration in the example given could be explained in a simple manner, if it is assumed that the sorption of cesium is controlled by the strongly acting cesium in liquid phase and not by the concentration of the other competing cations. The ion exchange capacity is not a satisfactory parameter for characterising the retention power of a material ; the chemical concentration of the cations to be removed should also be included.

P. BO, Denmark

Thank you for your suggestion that the cesium distribution might depend mainly on cesium concentration and less on the other ions. This is clearly what the data show. What we are looking for is an additional mechanism - besides ion exchange - that will explain this. Such a mechanism could possibly involve sorption sites that have a very high affinity for cesium as compared to other ions.

L.R. DOLE, United States

A similar study is in its third year at ORNL, under the supervision of J. Johnson, B. Meyer and K. Kraus. Two important differences are : the use of monomineralic samples and the rigorous sample purification with the Jackson procedure. In general, the results show :

1) In many cases these purified materials show classic ion exchange behaviour with widely varying ionic strengths and tracer concentrations. This shows that with extreme treatment a natural clay system may be forced to behave classically.

2) The Kd's produced are generally conservative, in that they are lower than for the more complex natural matrices.

3) The studies show that in some cases the results of tests performed with highly concentrated solutions are valid for many systems.

P. BO, Denmark

Thank you for your comments and information on the result of the work of Johnson, Meyer and Kraus.

We have not worked with extremely purified clay systems, as Johnson et al., but we have selected clays formed predominantly by montmorillonite, illite or kaolinite. I expect that when we extend the investigations from the kaolinite to the montmorillonite we will also find properties that are in better agreement will classical ion exchange.

THE CHARACTERIZATION OF CARBONATE COMPLEXES,
WHICH MAY BE FORMED IN GROUNDWATER

K. Nilsson
Risø National Laboratory
DK-4000 Roskilde, Denmark

A study of bicarbonate and carbonate complexes with selected
nuclides in radioactive waste is under progress and the current
status is reported.

Complex formation between cationic radionuclides and anions present in ground water may lead to a decrease in the over-all distribution coefficient of radionuclides between soil and ground water enhancing greatly migration of complexes with the ground water. Among the important complex-forming anions in ground water are the chloride, the fluoride, the sulphate and the carbonate/bicarbonate ions. The physico-chemical constants relevant for describing the formation of complexes between radionuclides and the chloride, fluoride and sulphate ions are known to the extent, that their influence on migration phenomena may be estimated, but experimental data concerning the stability of carbonate complexes with the long-lived nuclides of radioactive waste is sorely missing.

Very little reliable data exist because carbonate and bicarbonate compounds are generally insoluble, and as a consequence experimental work is hampered by precipitations. In addition, the reversible equilibrium, carbon dioxide and water forming carbonic acid, require that experiments be carried out under constant carbon dioxide pressure.

At the Risø National Laboratory in Denmark, we have undertaken to investigate the stability of carbonate/bicarbonate complexes with selected radionuclides. Our method of choice is the method of liquid-liquid partition. In this method, two immiscible solvents containing a complexing ion and the radionuclide to be studied, are mixed with either air or carbon dioxide in the gas phase, until complete equilibration. The distribution of the radionuclide between the two solvents is measured as a function of pH. The data thus obtained, is correlated with the theoretical model, and the desired formation constants obtained. It is a virtue of this method, that the radionuclide is used in tracer amounts. Then each complex contain only one metalion making theoretical treatment easier, and of course, the solubility problems are minimized. The method is basically a competition method in that a ligand, which is well suited for complexing with the radionuclides of interest, competes with carbonate and bicarbonate ions.

When the radionuclide, M and the complexing agent, A are mixed, a series of complexes, MA, MA_2, MA_3, ... may be formed in a stepwise fashion. Each step is characterized by an equilibrium constant, K_1, K_2, K_3, ..., or more commonly, the equilibrium is expressed as a function of the starting materials, M and A and the formation constants, β_1, β_2, β_3, ..., where

$$\beta_n = \prod_0^N K_n.$$

$$M + A \rightleftharpoons MA \qquad [MA] = K_1[A][M] = \beta_1[A][M]$$

$$MA + A \rightleftharpoons MA_2 \qquad [MA_2] = K_2[A][MA] = \beta_2[A]^2[M]$$

$$MA_2 + A \rightleftharpoons MA_3 \qquad [MA_3] = K_3[A][MA_2] = \beta_3[A]^3[M]$$

$$\vdots \qquad\qquad \vdots \qquad\qquad \vdots$$

The total concentration of the radionuclide, M in the system is the sum of the free metalion and the various complexes, all carrying different charges of which one of the charges may be zero.

$$C_M = M^{+m} + MA^{+(m-1)} + MA_2^{+(m-2)} + MA_3^{+(m-3)} + \cdots\cdots\cdots$$

In a two-phase liquid system the uncharged complex, for example, MA_c will distribute itself between the two phases, whereas all the charged molecules will remain in the aquous phase. $(K_D)_M$ is the distribution coefficient of the radionuclide.

$$(K_D)_M = \frac{[MA_c]^{organic}}{([M] + [MA] + [MA_2] + [MA_3] + \ldots)^{aqueous}} = \frac{[MA_c]^{org}}{(\sum_0^N [MA_n])^{aq}}$$

$(K_D)_M$ can also be expressed solely by the concentration of the complexing ion and the various stability constants. λ is the distribution coefficient of the uncharged molecule between the organic and aqueous phase.

$$(K_D)_M = \frac{\lambda \beta_c [A]^c}{1 + \beta_1 [A] + \beta_2 [A]^2 + \beta_3 [A]^3 + \ldots} = \frac{\lambda \beta_c [A]^c}{\sum_0^N \beta_n [A]^n} \qquad (1)$$

Since $(K_D)_M$ often varies over several decades it is common to use the logarithmic expression for graphic presentation.

$$\log (K_D)_M = \log \lambda \beta_c + c \log [A] - \log \sum_0^N (\beta_n [A]^n)$$

At low concentrations of the complexing ion, [A] there is very little complex formation. The last term more or less disappears and one is left with an initial straight-line relationship the slope of which indicate the number of ligands in the uncharged complex, if $\log (K_D)_M$ is plotted against $\log [A]$. Then as the concentration of [A] increases, there is a leveling off to a horisontal line, when the metalion is present solely as the uncharged complex and

$$\log (K_D)_M = \log \lambda$$

As pH is increased above neutrality there exist the possibility of hydroxide- and mixed hydroxide formation which equation (1) should take into account. Thus with air as the gas phase the following equation is valid,

$$(K_D)_M = \frac{\lambda \beta_c [A]^c}{\sum_0^N \beta_n [A]^n + \sum_0^I \gamma_i [OH]^i + \sum_0^J \sum_0^R \epsilon_{jr} [OH]^j [A]^r} \qquad (2)$$

When CO_2 is present, it will react with water forming carbonic acid which will then dissociate forming bicarbonate and carbonate. The concentration of each of these species is proportional to the CO_2 pressure,

$$CO_2 + H_2O \rightleftharpoons H_2CO_3; \qquad [H_2CO_3] = 10^{-1.47} \, pCO_2$$

$$H_2CO_3 \underset{}{\overset{pK\ =\ 6.4}{\rightleftharpoons}} H^+ + HCO_3^-; \qquad [HCO_3^-] = (10^{-7.87}/[H^+]) \, pCO_2$$

$$HCO_3^- \underset{}{\overset{pK\ =\ 10.3}{\rightleftharpoons}} H^+ + CO_3^{--}; \qquad [CO_3^{--}] = (10^{-18.17}/[H^+]^2) \, pCO_2$$

and an extra term is added in the expression for the distribution coefficient.

$$(K_D)_M^{CO_2} = \frac{\lambda \beta_c [A]^c}{\overset{N}{\underset{0}{\Sigma}} \beta_n [A]^n + \overset{I}{\underset{0}{\Sigma}} \gamma_i [OH]^i + \overset{JR}{\underset{00}{\Sigma\Sigma}} \epsilon_{jr} [OH]^j [A]^r + \overset{K}{\underset{0}{\Sigma}} \delta_k [CO_2]^k} \tag{3}$$

Combining expressions (2) and (3) results in expression (4).

$$\frac{(K_D)_M - (K_D)_M^{CO_2}}{(K_D)_M^{CO_2}} = \frac{\overset{K}{\underset{0}{\Sigma}} \delta_k [CO_2]^k}{\overset{N}{\underset{0}{\Sigma}} \beta_n [A]^n + \overset{I}{\underset{0}{\Sigma}} \gamma_i [OH]^i + \overset{JR}{\underset{00}{\Sigma\Sigma}} \epsilon_{jr} [OH]^j [A]^r} \tag{4}$$

which will be used to fit the experimental data in the attempt to determine the formation constants of the bicarbonate/carbonate complexes.

In the experiments, 4 ml of an organic phase containing the complexing ion is mixed with 4 ml of an aqueous phase containing the radionuclide in 0.1 M KNO_3, pH being adjusted with NaOH. The mixture and the air space above is saturated with the gas required (CO_2 or air or CO_2/N_2 mixtures), after which the test tubes are closed, and then placed on a wheel rotating slowly at 35 rpm in a plastic glovebox containing the same gas as used for saturation. The equilibration is allowed to go on for 17 hours. After centrifugation, 3 ml of the upper phase is transferred using a disposable pipette to another plastic tube. The sample is counted in a well-type NaI scintillation counter. The lower phase is counted as well and finally the pH of the equilibration is measured.

The distribution coefficient of the electroneutral complex is normally calculated solely on data obtained from the organic phase using the following equation,

$$(K_D)_M = cpm(organic\ phase)/(cpm(total) - cpm(organic\ phase))$$

Xylene or isoamylalcohol or mixtures of these is used as the organic phase. In order to ensure a constant ionic strength, 0.1 M KNO_3 is used as the aqueous phase, which also contains the radionuclide, Europium(154) or Strontium(85).

1-phenyl-3-methyl-4-acyl-pyrazolone-5, which is a diketone well suited for complexing with the radionuclides of interest in this study, is employed as the complexing agent.

$$R = -\underset{O}{\overset{\parallel}{C}}-CH_3 \quad \text{(acetyl)} \quad \text{or} \quad -\underset{O}{\overset{\parallel}{C}}-C_6H_5 \quad \text{(benzoyl)}$$

The compound is a weak acid with a pK of 3.99. In a two-phase liquid system, the titration curve is shifted toward higher pH's as shown in figure 1. The size of the shift is dependant on the solubility of the pyrazolone in the organic phase. The value of the appeerent pK (= pK') must be determined for each pyrazolone derivative in each two-phase system employed, since pK' is necessary for calculating the concentration of the complexing specie, the pyrazolone anion, at various pH's.

To summarize; in the experiments $(K_D)_M$ and $(K_D)_M^{CO_2}$ are determined as a function of pH. From the pH, the concentration of the ligand [A] can be determined in addition to the concentration of [OH]. From the CO_2 pressure and the pH, the concentration of carbonate/bicarbonate ions is determined. These data are the experimental data required for the theoretical equations.

Figure 2 show the result obtained with Europium as the radionuclide and acetylpyrazolone as the complexing ion with air in the gas phase. As pH increases there is a leveling off followed by a sharp drop starting at pH 6, which is due to the formation of insoluble $Eu(OH)_3$. Taking hydroxide formation into account and then fitting the data using equation (2), the result shown in figure 3, which also gives the constants, is obtained.

The result obtained with Europium and benzoylpyrazolone and air in the gas phase is shown in figure 4, but this time, equation (2) can not be fitted to the data on the assumption. that insoluble hydroxide is present. Probably a mixed complex is formed the exact composition of which, is still being investigated.

When CO_2 is present in the Europium-benzoylpyrazolone system there is a marked effect as shown in figure 5. Although the result has not yet been evaluated quatitatively, it is evident that strong bicarbonate/carbonate complexes are formed.

CO_2 also has an effect on the Strontium-benzoylpyrazolone system as can be seen in figure 6, but Strontium carbonate complexes are probably of only moderate stability.

This investigation is still in progress and continues with the purpose of determining the stability constants of carbonate/bicarbonate complexes of selected radionuclides.

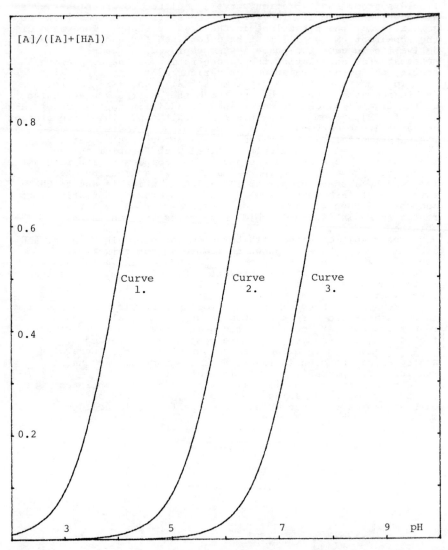

Figure 1: Curve 1: Titration of Acetyl- or Benzoylpyrazolone in
           aqueous solution (pK = 3.99), Curve 2: Titration of
           Acetylpyrazolone in Xylene/Water (pK' = 6.02),
           Curve 3: Titration of Benzoylpyrazolone in Xylene/
           Water(pK' = 7.43).

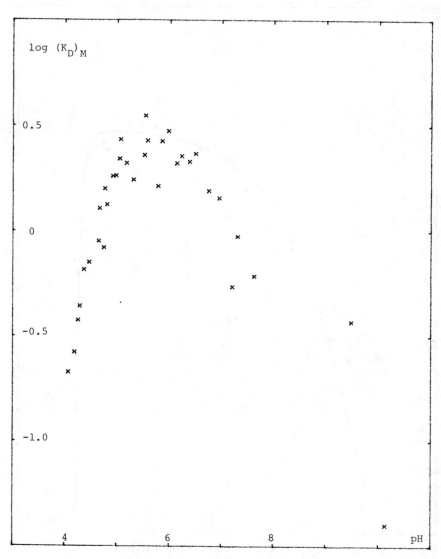

Figure 2:   log $(K_D)_M$ as a function of pH.
Europium/Acetylpyrazolone in Xylene/Water.
Saturation with Air.

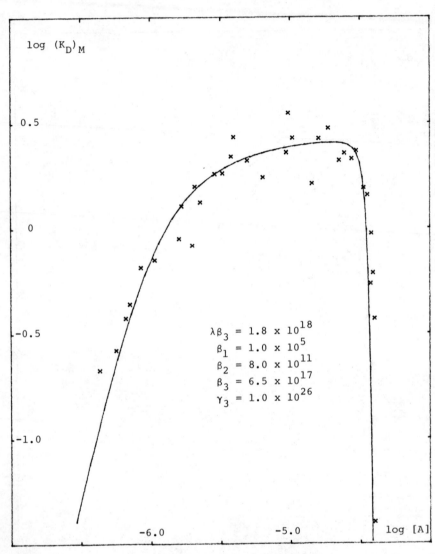

Figure 3:   log $(K_D)_M$ as a function of log [A].
Europium/Acetylpyrazolone in Xylene/Water.
Saturation with Air.

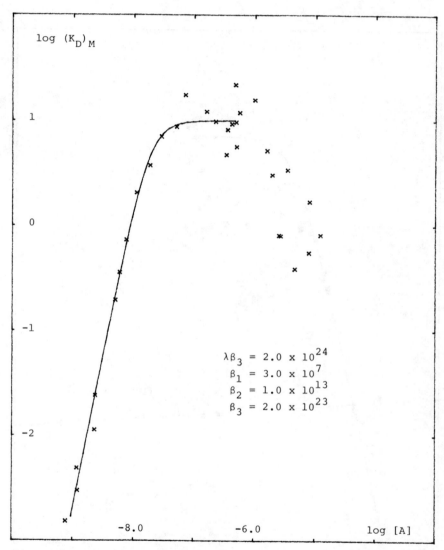

Figure 4:   log $(K_D)_M$ as a function of log [A].
Europium/Benzoylpyrazolone in Xylene/Water.
Saturation with Air.

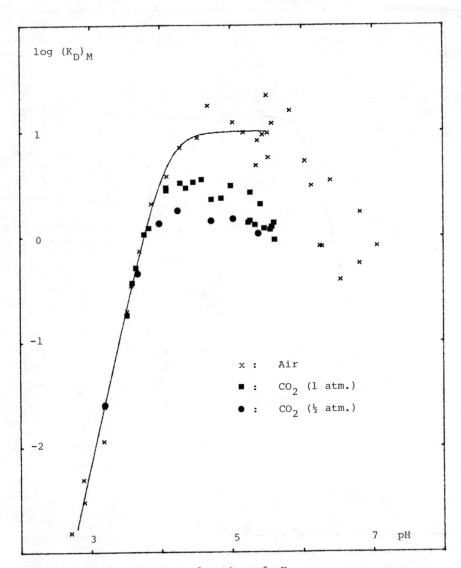

Figure 5:   log $(K_D)_M$ as a function of pH.
Europium/Benzoylpyrazolone in Xylene/Water.

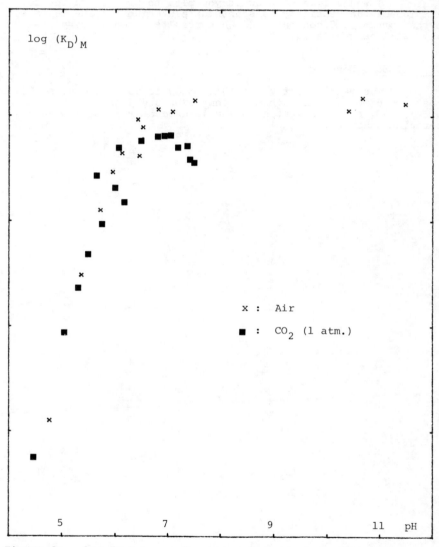

Figure 6:   log $(K_D)_M$ as a function of pH.
            Strontium/Benzoylpyrazolone in Isoamylalcohol/Water.

Discussion

F. GIRARDI, CEC

I would like to know it you plan to proceed to the determination of the stability constants of plutonium and americium within the current CEC contract and when results, even if very preliminary, will be available.

K.K. NILSSON, Denmark

The main objective of this study is to determine stability constants of plutonium carbonate complexes and this will be done during 1979. At the moment europium is used as representative of americium behavior.

RETENTION DES RADIONUCLEIDES A VIE LONGUE
PAR DIVERS MATERIAUX NATURELS

D. Rançon *
J. Rochon * *
* CEA, Institut de Protection et de Sûrete Nucléaire, CEN Cadarache
** BRGM, département MGA, Orléans

Résumé . Dans le cadre des études de stockages des déchets radioactifs à grande profondeur, on a recherché des matériaux susceptibles de constituer des barrières géochimiques pour les Transuraniens (Np, Pu, Am) et les produits de fission à vie longue (Sr, Zr, Cs, Sm, Tc, I). Les silico-aluminates retiennent, en général assez bien les transuraniens et les produits de fission cationiques; parmi eux l'attapulgite, bentonite et l'illite, très efficaces, constitueraient d'excellentes barrières géochimiques. Mais ils ne retiennent pratiquement pas les P.F. anioniques (Tc, I), ces derniers peuvent être piégés selon un processus différent par des minerais de Cu, Pb ou Fe (Chalcopyrite, galène, sidérose).

1.   INTRODUCTION

Dans le cadre du projet de stockage des déchets radioactifs dans les massifs cristallins en grande profondeur, on a entrepris la recherche des "barrières géochimiques", matériaux destinés à faire écran entre le produit stocké et la roche encaissante pour empêcher ou limiter la migration d'éléments radio-actifs sous l'effet de courants d'eau éventuels. Dans cette étude, on présente une sélection de sorbants, aptes à retenir les éléments transuraniens et les produits de fission à vie longue que pourraient contenir les déchets.

2.   DEFINITION DE LA RETENTION ET METHODES DE MESURE

Pour définir la rétention par un matériau solide d'un corps en solution, le coefficient de distribution ou Kd, rapport des concentrations à l'équilibre du corps en phase solide et en phase liquide, est le terme le plus utilisé et le plus pratique.

Le Kd englobe tous les mécanismes physico-chimiques susceptibles de se produire lors du passage en phase solide d'un élément en solution (échange d'ions ad. ou absorption, précipitation) ; aussi les valeurs du Kd ne sont pas suffisantes pour caractériser les phénomènes de sorption, mais elles sont très utiles lorsqu'on veut obtenir des valeurs comparatives de la rétention d'un corps par divers matériaux, ce qui était le cas de cette étude.

Dans cette première étape, le Kd a été mesuré par la méthode statique en vase clos [1] . On a opéré à température ambiante en utilisant 1g de matériau dans 20cm3 de solution marquée. On a utilisé une eau dont la minéralisation représentait en moyenne celle des eaux des massifs granitiques.

Parmi les facteurs physiques et chimiques qui influent sur la rétention [1] , le pH joue un rôle essentiel car souvent de faibles variations de pH engendrent de grandes variations du Kd quels que soient les mécanismes de réten-tion. C'est pourquoi tous les systèmes radioéléments-solide-liquide ont été considérés à divers pH.

3.   LES RADIOELEMENTS CONSIDERES

Les radioéléments de longue période dont la teneur initiale est suffisamment importante pour constituer un danger à long terme (après 100 ans et 1000 ans de stockage) sont en nombre limité, ils sont présentés dans le tableau I

Tableau I

Les radioéléments étudiés

Catégorie	Elément	Isotopes à vie longue (T an)	Isotopes utilisés T
Trans-uraniens	Np Pu Am	237 , 239 239 , 240, 242 241 , 243	237 $(2,210^6$ a) 239 $(2,4\ 10^4$a) 241 (470 a)
Produits de fission cationiques	Sr Zr Cs Sm	90 (28) 93 $(9,5\ 10^5)$ 137 (30) 135 $(2,9\ 10^6)$ 151 (90)	85 (64j) 95 (65j) 134 (2a) 153 (47 h)
P.F. anioniques	Tc I	99 $(2,1\ 10^5)$ 129 $(1,6\ 10^7)$	96 (4,3 j) 131 (8 j)

Après 1000 ans de stockage subsisteraient en quantité notable les trans-uraniens et parmi le P.F., compte tenu des rendements de fission, essentiellement le 93 Zr et le 99 Tc.

La production d'I-129 est très faible, environ 10^{-7} fois l'activité totale des déchets après 10 ans de décroissance, et sa présence dans des déchets tels que les produits vitrifiés est improbable. On l'a toutefois pris en considération en raison de son danger biologique et pour le cas où il serait stocké à part.

4.   LES MATERIAUX ETUDIES (PREMIERE SELECTION)

    Après une première élimination on a retenu des silico-aluminates présentés dans le tableau II.

Tableau II
Les silico-aluminates étudiés

Nom	Nature	Symbole	Référence
Attapulgite	Argile fibreuse	A	2
Bauxite	Mélanges d'hydrates d'Al et Fe et d'argiles		
Bentonite	Argile de la famille montmorillonite	Ba	3
Clinoptilolite	Zéolite naturelle	Be	2
Illite	Argile phylliteuse	C	3
Kaolinite	Argile phylliteuse	I	2
Sépiolite	Argile fibreuse	K	2
Vermiculite	Argile phylliteuse	S	2
		V	2

5.   RESULTATS, VALEURS DES Kd

5.1 Kd au pH d'équilibre

    Dans l'eau (pH 6,9) le matériau libère des ions H^+ ou OH^- et l'ensemble acquiert un pH d'équilibre variant de pH 6,5 (kaolinite) à pH 10 (bentonite)

    Les Kd des divers systèmes (élément-eau-matériau) au pH d'équilibre sont indiqués sur la figure 1.

6.   INTERPRETATION, SELECTION

    On attribue à chaque couple radioélément matériau, un "indice de rétention" correspondant à la caractéristique du logarithme décimal du Kd. Les indices correspondants aux Kd mesurés au pH d'équilibre sont reportés sur le tableau III. Les sommes de ces indices permettent de comparer les propriétés sorbantes des divers matériaux et les propriétés globales des radioéléments.

Tableau III
Indices de rétention au pH d'équilibre

pH	7,5	7	10	6,5	7,5	6,5	8,5	8,5	Σi
mat élt	A	Ba	Be	C	I	K	S	V	(élément)
S_r	2	1	3	3	3	2	2	2	18
Zr	5	6	5	4	5	3	5	3	36
Cs	3	2	3	4	3	2	4	4	25
Sm	5	6	5	4	5	4	5	4	38
Np	1	2	4	1	1	1	1	1	13
Pu	5	4	4	3	4	3	5	5	32
Am	4	4	4	4	4	3	4	4	31
Σi (mat)	25	25	28	23	25	18	26	23	/
Tc	0	1	0	1	0	1	0	0	3
I	0	0	0	0	0	0	0	0	0

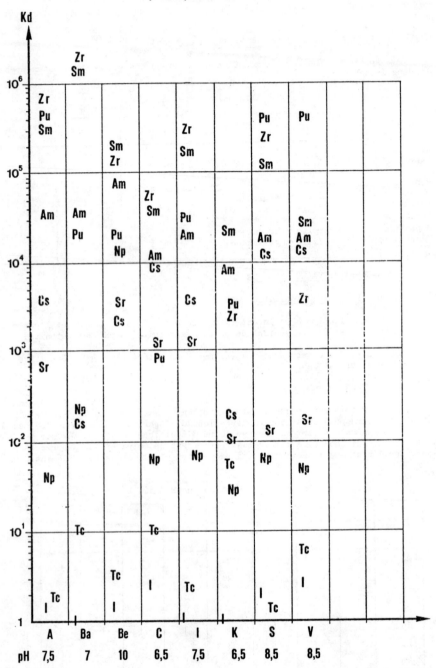

Figure 1.   Kd au pH d'équilibre

6.1 Comportement des radioéléments

L'examen de la figure 1 et du tableau III permet de séparer immédiatement 2 catégories de radioéléments : les cations (P.F. et TU) d'une part et les P.F. anioniques d'autre part. En effet, le technetium (Tc O_4^- en solution) et l'iode (I^- en solution) ne sont pas ou très peu retenus par les silico-aluminates; il faudra donc trouver pour eux des barrières géochimiques d'un autre type.

Dans le cas des cations, on constate que 4 éléments, Zr, Sm, Pu, Am, sont très fortement retenus par tous les matériaux, les Kd toujours supérieurs à 1000 dépassent souvent 10^5 et même 10^6 cm3.g^{-1}; ce sont les éléments chez qui les phénomènes de précipitation d'hydroxydes sont prépondérants aux valeurs de pH supérieures à 2 ou 3. Le Cs est dans l'ensemble bien retenu et le Sr surtout par la bentonite, la clinoptilolite et l'illite. Dans le cas du Np seule la bentonite assure une bonne rétention de cet élément qui se trouve en solution sous la forme Np O_2^+.

6.2 Classements des barrières géochimiques au pH d'équilibre

La somme des indices de rétention représentée sur le tableau III permet de faire le classement suivant :
1 bentonite (28)
2 attapulgite, bauxite, illite, sépiolite (25 et 26)
3 clinoptilolite, vermiculite (23)
4 kaolinite (18)

6.3 Sensibilité aux variations de pH

Sur chaque ensemble, on a mesuré les variations du Kd en fonction du pH; il en est résulté 106 courbes Kd = (pH). Les résultats ont été regroupés dans 4 zones de pH (5 à 6, 6 à 7, 7 à 8 et 8 à 9). La somme des indices de rétention de chaque ensemble élément-matériau dans ces 4 zones de pH sont indiqués dans le tableau IV. et les Kd sur les figures 2 à 5.

Tableau IV
Σi (matériau) dans les zones de pH

	A	Ba	Be	C	I	K	S	V
5 < pH < 6	23	18	18	21	21	15	21	21
6 < pH < 7	23	23	19	22	25	18	20	21
7 < pH < 8	25	25	20	20	27	22	23	25
8 < pH < 9	26	26	24	21	27	26	25	23
Total	97	92	81	83	100	80	89	90

Le total des Σi dans les 4 zones de pH montre que l'illite et l'attapulgite sont, parmi ces divers matériaux, les moins sensibles aux variations de pH.

7. SELECTION DE BARRIERES GEOCHIMIQUES VIS A VIS DES CATIONS

Parmi les matériaux étudiés; l'attapulgite, la bentonite et l'illite, sont ceux qui assurent la meilleure rétention des éléments transuraniens et des produits de fission cationiques, comme ces argiles naturelles existent en abondance dans divers gisements européens, elles pourraient être choisies pour constituer de bonnes barrières géochimiques artificielles ; de plus la bentonite, outre qu'elle est la seule à retenir correctement le neptunium, possède des propriétés physiques intéressantes (gonflement, gélification) qui la font utiliser à l'étanchéification d'ouvrages de génie civil et au colmatage des fissures.

La bauxite serait aussi intéressante mais elle retient mal le Sr et le Cs. La sépolite, en dépit de ses bonnes propriétés sorbantes, n'est pas à conseiller, car elle se décompose en milieu acide en dessous de pH 5. La clinoptilolite et la vermiculite ont un Σi inférieur au pH d'équilibre à celui des matériaux précédents et de plus elles ne sont pas actuellement disponibles en quantités industrielles dans des gisements européens. La kaolinite, en raison de son Σi relativement faible n'est pas recommandée.

8. CAS DE L'IODE ET DU TECHNETIUM

Les silico-aluminates arrêtant les cations ayant révélé leur incapacité à retenir les radioéléments anioniques, on a recherché d'autres minéraux à partir de 3 objectifs :

- Probabilité de forte densité de sites électropositifs sur lesquels pourraient se fixer les anions (hydroxyde de Fe ou d'Al)
- Possibilité de réaction d'oxydo-réduction entre la solution et le matériau (sels ferreux)
- Possibilité d'apparition dans la solution de cations pouvant former avec l'anion des composés insolubles (sels de Cu, Pb, Hg, Ag dans le cas de l'iode).

On a donc étudié un certain nombre de minerais assez fréquents dans la nature pour être obtenus sans frais excessifs (minerai d'Al, de Fe, de Cu et de Pb) (Tableau V, réf. 3).

Les valeurs du Kd de l'iode - 131 et du Tc 99 entre la solution et ces divers matériaux sont indiqués sur le tableau V.

Dans le cas de l'iode, les valeurs de Kd montrent que la rétention de l'ion iodure proviennent plus de la formation d'iodures insolubles que de l'absorption de I^- sur des sites électro-positifs. Parmi les minerais de cuivre et de plomb testés, la chalcopyrite donne les meilleurs résultats, une étude plus approfondie de ce minéral a montré que le Kd augmente très rapidement quand le pH diminue (entre 1000 et 3000 $cm^3.g^{-1}$ de pH 6 à 5.

Tableau V
Kd du Tc 99 et de I 131 entre la solution et divers minerais

Minerai	Composition	Kd $cm^3.g^{-1}$		pH
		$Tc\ O_4^-$	I^-	
Bauxite	hydrates d'alumine	0	1	7,5
Latérite	hydroxydes de Fe et d'Al	0	0,2	7,5
Limonite	hydroxydes de Fe	6	6	6,7
Goethite	FeO . OH	9	1	7,7
Sidérite	Fe CO_3	60	0,2	7,3
Chalcopyrite	Fe Cu S_2	0,4	215	7,8
Chrysocolle	Cu SiO_3, nH_2O	-	10	7,2
Cérusite	Pb CO_3	2	3 à 8	7,7
Galène	Pb S	40 à 3000	15 à 50	7,5 à 7

Dans le cas du technetium, la sidérite et la galène donnent les meilleurs résultats, le Kd avec la galène augmentent très rapidement quand le pH diminue (de 4 à plus de 10^5 de pH 9 à 6).

La rétention du Tc O_4^- par la galène pourrait être due à une réduction du Tc O_4^- en Tc O_2 insoluble par les ions sulfures. Dans le cas de la sidérite ce serait plutôt une absorption sur des colloïdes d'hydroxydes de Fer. Ces hypothèses sur les mécanismes de rétention du Tc devraient être vérifiées par une étude plus précise.

9.   CONCLUSION

Les minéraux argileux, par leurs propriétés physiques, se prêtent bien aux diverses opérations concernant la sûreté des stockages (conditionnement, étanchéité); d'autre part, ils constituent aussi de bonnes barrières géochimiques vis à vis de la plupart des radioéléments présents dans les déchets, l'expérience a montré que l'attapulgite, la bentonite et l'illite étaient les plus efficaces. Toutefois ces matériaux ne retiennent pas les radioéléments anioniques (iode et technetium) qui peuvent être retenus par des minéraux de tout autre nature (minerais de Cu, Pb, Fe). La barrière artificielle idéale devrait donc comporter une barrière aux cations suivie d'une barrière aux anions en aval du courant d'eau présumé et non l'inverse pour éviter la saturation des sites de sorption dans les argiles.

A la suite de ces essais, il est nécessaire pour une qualification à long terme d'effectuer des essais dynamiques sur une plus grande échelle afin d'évaluer notament, les phénomènes de reversibilité. Des expériences effectuées sur petites colonnes sont actuellement développées à l'échelle pilote sur grandes colonnes.

Références

[1] Rancon R. "L'aspect géochimique pour les évaluations des transferts des radioéléments dans les milieux poreux souterrains" Rapport CEA-R-4937 (1978)

[2] Caillere S., Henin S. "Minéralogie des argiles" Masson ed. Paris (1963)

[3] Aubert G., Guillemin C., Pierrot R. "Précis de minéralogie" Masson - BRGM ed Paris (1978)

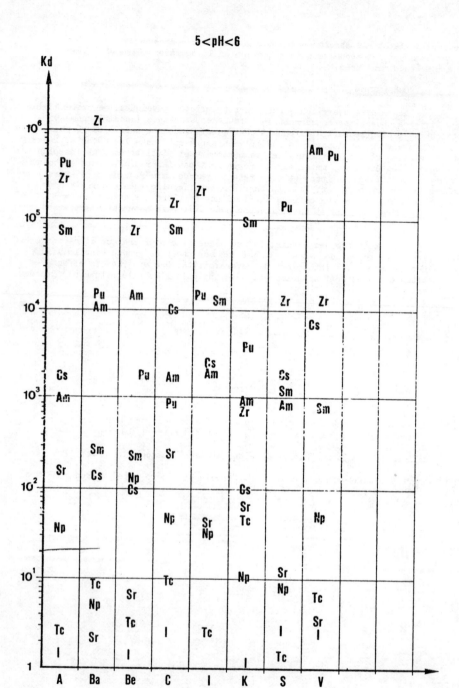

Figure 2. Kd moyens entre pH 5 et 6

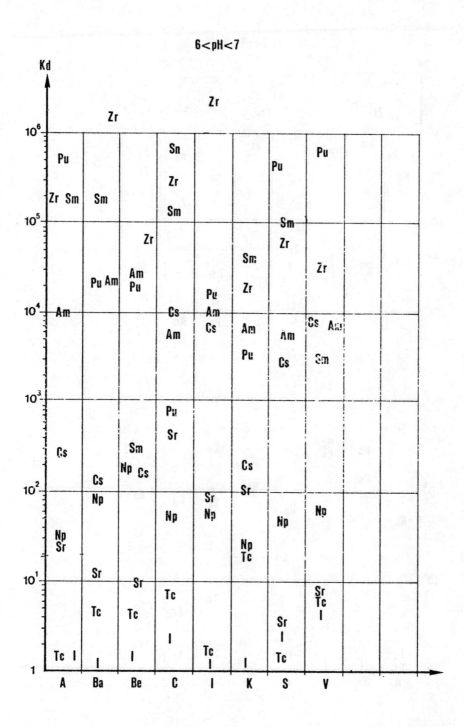

Figure 3.   Kd moyens entre pH 6 et 7

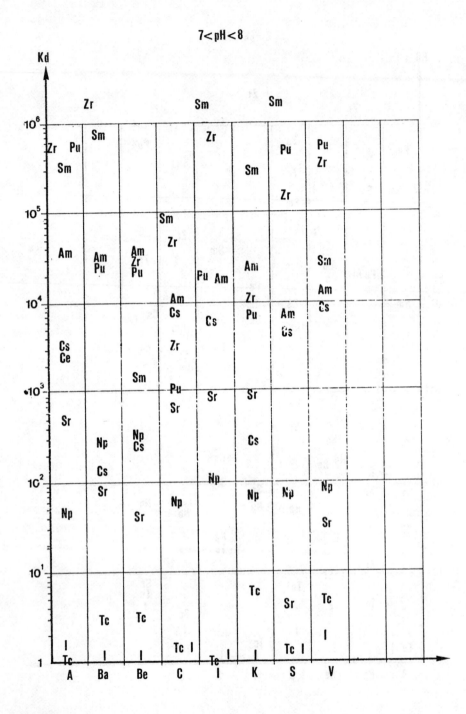

Figure 4. Kd moyens entre pH 7 et 8

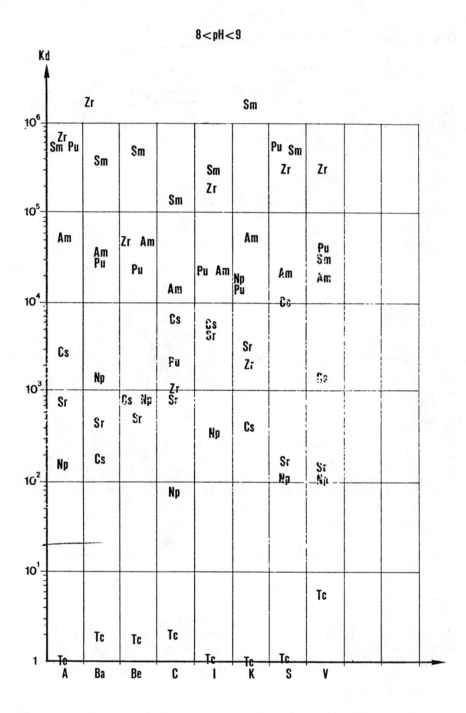

Figure 5. Kd moyens entre pH 8 et 9

Discussion

F. van DORP, The Netherlands

Material put around radioactive waste to adsorb radio-nuclides leached from the waste, for example in glass form, should be less leachable, or soluble than the waste itself, otherwise this material will disappear before the waste. Have you any idea about the stability against dissolution of the clay minerals and salts discussed in your paper ?

D.L. RANCON, France

Nous n'avons pas encore fait d'expériences sur la disso-lution des silicoaluminates.

G. LENZI, Italy

Quel est le temps de contact à l'équilibre entre la solu-tion et les minéraux que vous avez essayés ?

D.L. RANCON, France

Pour chaque ensemble, on a effectué une évaluation de l'évolution du Kd en fonction du temps. Pour les produits de fission, l'équilibre est atteint en quelques heures. On a choisi un temps de contact de 48 h pour travailler dans les mêmes conditions dans le cas des produits de fission et des transuraniens pour lesquels la mise à l'équilibre est plus longue.

G. LENZI, Italy

J'ai également une remarque à faire sur les minéraux que vous avez décrits ici : il y a d'autres alumosilicates naturels, tels que le chabazite et la phillipsite, qui sont très répandus dans certains pays, et qui montrent une très bonne capacité d'échange pour les produits de fission, comme le cesium et strontium.

D.L. RANCON, France

Oui, il existe sûrement d'autres matériaux possédant de bonnes qualités sorbantes, l'heulandite par exemple ou diverses zeolites synthétiques.

G. LENZI, Italy

Ma troisième question est la suivante : quelle technologie pensez-vous utiliser pour la mise en place de matériaux naturels dans le terrain, pour obtenir les barrières géochimiques ? Par injection en suspension par exemple ?

D.L. RANCON, France

On n'a pas encore considéré la technologie de la mise en place des barrières. Peut-être nos collègues suédois auraient des informations sur ce sujet.

P. COHEN, France

En vue d'étendre le champ de l'étude des Kd, je voudrais faire les suggestions suivantes :

- Il faudrait tenir compte, dans la composition chimique de l'eau en contact avec la barrière géochimique, des ions qui vont passer en solution après corrosion de la première barrière, c'est-à-dire le conteneur.

- L'interaction entre la solution - modifiée ou non par la présence de produits de corrosion - et la barrière géochimique peut modifier la composition chimique de l'eau qui sera finalement en contact avec la "barrière géologique". Dans ce cas, ne serait-il pas intéressant d'examiner la modification du Kd sur la barrière géologique ?

D.L. RANCON, France

Nous nous sommes occupés des barrières secondaires et tertiaires. Pour tenir compte de votre proposition, il faudrait connaître la nature de la barrière primaire.

En ce qui concerne votre deuxième point, c'est précisément l'étude de l'échange d'ions que nous avons commencée pour interpréter les résultats obtenus.

A.A. BONNE, Belgium

Il y a encore une substance importante qu'on ne peut pas oublier dans ce problématique et qui, elle aussi a une interaction avec le milieu géologique. Il s'agit notamment des matériaux de structures, par exemple béton ou fonte. Ces matériaux peuvent présenter des volumes énormes dans certains types de formations-hôtes pour déchets radioactifs.

L.R. DOLE, United States

In the consideration of secondary engineering barriers, the analyses of Hill, in a previous paper, and Burkholder indicate that in many systems of interest, the nuclide retention of the barrier is not important. Perhaps, more important qualities of these barriers would be mechanical strength, impermeability to water, electrochemical buffering capacity, etc., which would hopefully increase the longevity of the waste canisters.

D.L. RANCON, France

Les diverses propriétés des barrières sont certainement très importantes, mais pourquoi ne pas y ajouter les qualités adsorbantes ?

C. MYTTENAERE, CEC

Si l'on s'en réfère aux travaux suédois décrits dans les rapports KBS, le récipient contenant les déchets sera entouré d'un manteau de bentonite. Cette façon de procéder confirme vos résultats en matière de rétention des formes cationiques.

D.L. RANCON, France

Oui, la bentonite est un des meilleurs matériaux testés. M. Rochon pourrait peut-être donner quelques renseignements complémentaires.

F.P. SARGENT, Canada

I think we can spend a few more minutes on this subject.

J. ROCHON, France

Je remercie Monsieur le Président de nous accorder quelques minutes supplémentaires pour présenter d'autres résultats effectués dans le cadre de ces travaux. Je vais le faire en commentant quelques figures.

1/ Les résultats présentés par Monsieur RANÇON décrivent la rétention des radioéléments par des minéraux pris à l'état brut. Or, lors d'un stockage définitif l'interaction élément-minéral ne se produira qu'après un temps important ; le minéral aura été lixivié et sera alors en équilibre avec l'eau du site. La figure 1 compare les K_d du césium avec de la vermiculite brute et avec de la vermiculite préalablement mise en équilibre avec l'eau des massifs granitiques ("vermiculite cristalline"). Il est visible que le traitement décationisant, subi par la vermiculite, affaiblit le phénomène de compétition ionique entre le césium et les autres cations, entraînant une rétention supérieure du césium par la "vermiculite cristalline". Les essais effectués avec les matériaux bruts vont donc dans le sens d'un accroissement de sécurité en minorant la rétention expérimentale.

2/ En ce qui concerne la rétention du technétium par la sidérite, la figure 2 montre qu'il ne s'agit pas d'une réduction de TcO_4^- en TcO_2 insoluble, puisque tous les points représentatifs des équilibres de la courbes $K_d=f(pH)$ sont dans le domaine du pertechnétate dans le diagramme $Eh=f(pH)$. Nous pensons qu'il s'agirait plutôt d'une rétention de TcO_4^- par des hydroxydes de fer.

3/ Les mesures de K_d englobent tous les phénomènes de rétention et constituent, de ce fait, des bons tests comparatifs des pouvoirs de sorption des minéraux. Toutefois, il est également intéressant de connaître les capacités maximales de rétention de ces minéraux pour un élément donné. C'est pourquoi nous avons déterminé des isothermes de rétention $S=f(C)$ (S = quantité d'élément fixé à l'équilibre par gramme de solide, C = quantité d'élément restant en solution par litre). Les figures 3 et 4 montrent les isothermes de rétention du

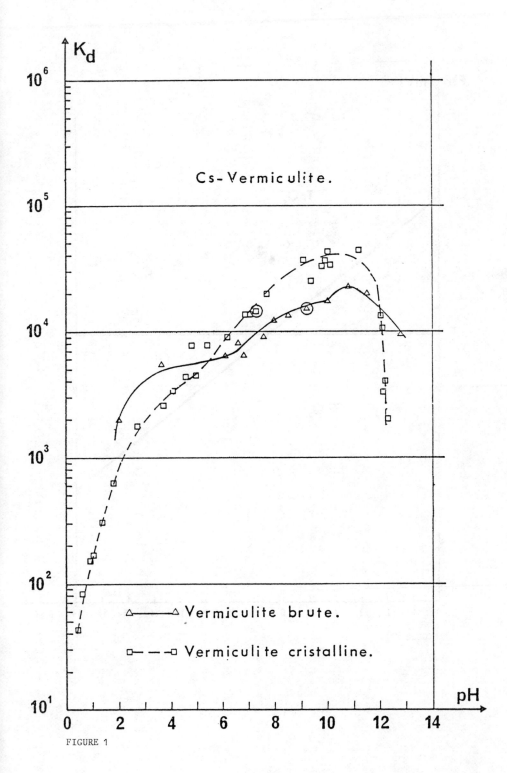

FIGURE 1

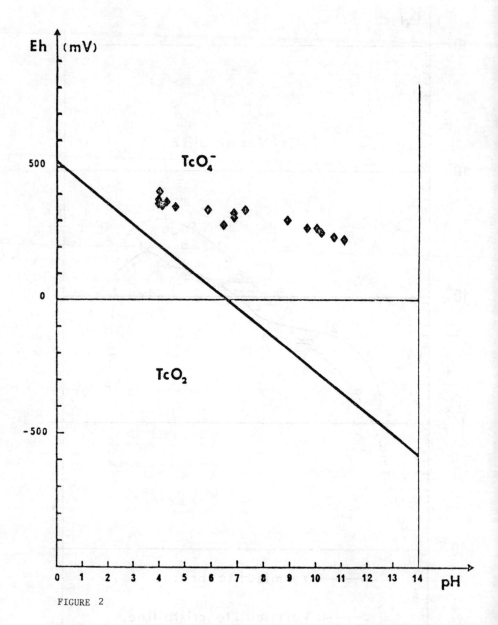

FIGURE 2

césium à 20°C et 50°C par l'illite et l'attapulgite. Notons que, dans le domaine considéré, la température a peu d'influence sur les courbes de capacité. Les capacités de rétention du césium sont, pour ces deux minéraux, de l'ordre de 50 meq/100 g.

4/ La sécurité des stockages est dépendante de la rétention définitive des radioéléments. C'est pourquoi nous avons effectué des tests de réversibilité en déterminant des isothermes de désorption : lorsqu'un équilibre de sorption est atteint, on ajoute à la suspension diverses quantités d'eau donnant d'autres états d'équilibre qui définissent l'isotherme de désorption issue de l'équilibre de sorption initial. Les formes obtenues pour ces courbes sont représentées sur la figure 5.

- Les isothermes du type ① sont caractéristiques d'un échange d'ions, d'une adsorption irréversible ou d'une combinaison de ces deux processus.

- les isothermes du type ② caractérisent des processus d'échange d'ions couplés ou non à une adsorption partiellement réversible.

- les isothermes du type ③ correspondent à un échange d'ions combiné ou non à une adsorption réversible avec hystérésis.

Ce type d'essais ne permet donc pas, par cette simple analyse, de conclure sur le mécanisme et la réversibilité des sorptions. D'autre part, cette méthode est limitée par le fait qu'on obtient rapidement des volumes de solution trop important pour la pratique expérimentale. C'est pourquoi, pour juger de la qualité des rétentions, nous avons dû recourir à des essais sur colonnes pendant lesquels on peut passer des volumes d'eau pratiquement illimités. Nous les présenterons ultérieurement [1]

5/ Nous avons toutefois tenté d'interpréter la sorption du césium à 20°C par la vermiculite comme un phénomène d'échange cationique.

Pour chaque équilibre, nous avons obtenu la distribution de toutes les espèces simples ou complexes en solution en introduisant les données analytiques à un programme de calcul des équilibres chimiques inspiré du "WATEQ" [2] . La vermiculite étant un échangeur essentiellement cationique, nous avons supposé que les ions et complexes anioniques (Cl^-, HCO_3^-, $NaSO_4^-$, $NaCO_3^-$ etc ...) s'en trouvaient exclus et que la pénétration des espèces non chargées ($CaCO_3°$, $MgSO_4°$, $CsCl°$, etc ...) était négligeable. Connaissant les ions contenus dans le minéral, avant interaction, nous avons pu calculer la constante

$$K = \frac{(\overline{Cs}).(\overline{H}).(Na^+).(K^+)}{(Cs^+)(H^+)(\overline{Na})\ (\overline{K})} \left[\frac{(Ca^{2+}).(Mg^{2+})}{(\overline{Ca})\ (\overline{Mg})} \right]^{1/2} \times \frac{T}{Q_0}$$

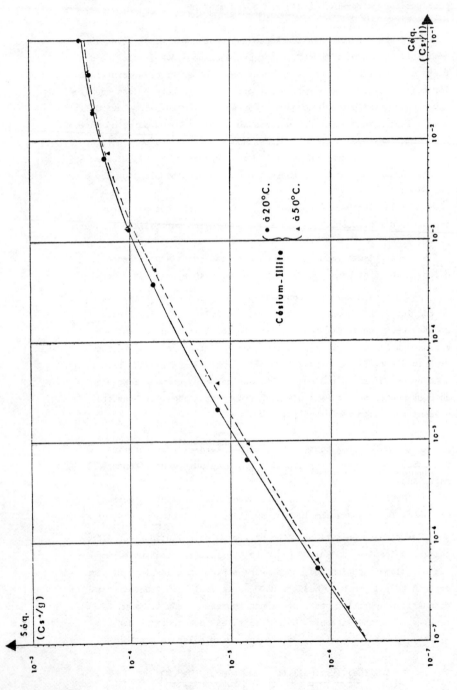

Séq.
(Cs⁺/g)

Céq.
(Cs⁺/l)

Césium-Illite $\left(\begin{array}{l} \bullet \text{ à 20°C.} \\ \blacktriangle \text{ à 50°C.} \end{array} \right.$

FIGURE 3

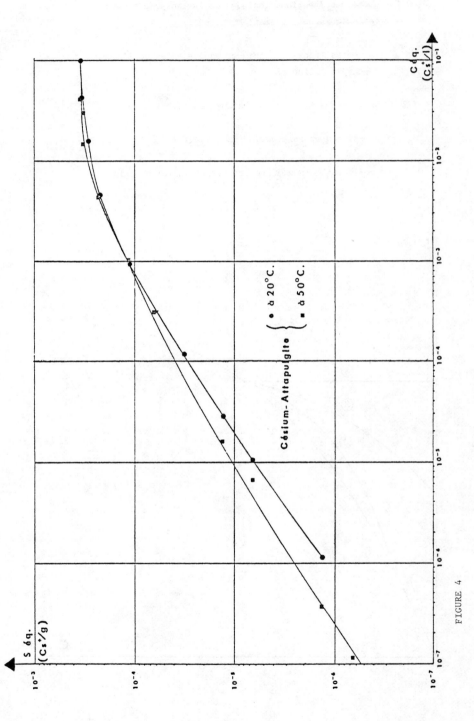

FIGURE 4

où

(M^{n+}) = somme des fractions ioniques équivalentes des espèces
cationiques de l'élément M en solution

(M) = somme des fractions ioniques équivalentes de l'élément
M dans le solide

T = somme des activités des cations en solution (eq/l)

Q$_0$ = capacité d'échange cationique (eq/g)

K = coefficient d'équilibre relative à la loi d'action de
masse

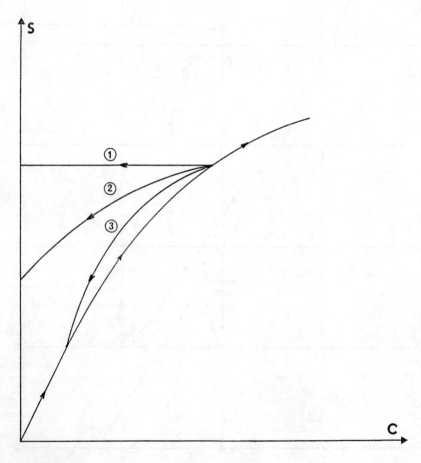

FIGURE 5

On conçoit que cette équation, même si elle montre que l'échange d'ions est dans ce cas le phénomène de rétention prépondérant, est inexploitable dans un cadre prévisionnel du transfert. C'est pourquoi nous avons simplifié le problème en cherchant une relation empirique à l'isotherme représentée en diagramme carré (donc en fraction ionique équivalente). (figure 6)

Cette relation Y = f(X) sera plus commode à introduire dans une équation représentative de la dynamique du transfert de l'élément dans le site du stockage.

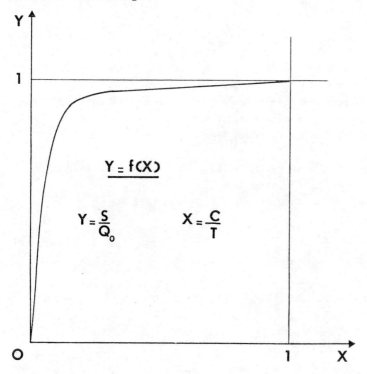

FIGURE 6

Notons que la fonction f est souvent telle, que l'on peut considérer que l'on a affaire à un échange de cations homovalents entre le cation considéré et l'ensemble des autres cations pris globalement.

$$Y = \frac{KX}{(K-1)X + 1} \iff K = \frac{S}{C} \frac{(T-C)}{(Q_o - S)}$$

Note :

Si C et S sont très petits vis-à-vis de T et Q_o

$$S = \frac{K \cdot Q_o}{T} \cdot C = K_d \cdot C$$

Nous voyons donc que l'on ne peut parler d'un K_d constant pour un minéral et un élément donnés que si la composition cationique de l'eau environnante est constante $(T=c^{ste})$.

BIBLIOGRAPHIE

[1] ROCHON J., RANÇON D., GOURMEL J.P., "Recherche en laboratoire sur la rétention et le transfert de produits de fission et de transuraniens en milieu poreux", Colloque International sur l'évacuation des déchets radioactifs dans le sol, IAEA - SM - 243/155, OTANIEMI, 2-6 juillet 1979

[2] TRUESDELL A.H., JONES B.F., "WATEQ : computer programme for calculing chemical equilibrium of natural waters", J. Res. U.S. Geol. Survey, Vol. 2, n° 2, (1974), pp. 233-246

GENERAL DISCUSSION

DISCUSSION GÉNÉRALE

F.P. SARGENT, Canada

I would like to thank you for your presentations. I think that we could spend a little more time on a few points which were raised by some previous speakers this morning. For example, just before we had our break, Les Dole, made some comments and I cut Peter Bo off. This might be the time for him to reply. I would urge him to make his reply reasonably short.

P. BO, Denmark

I was very glad for the comment about the work going on in the United States. It seems that they are trying to work with extremely pure systems. Most people working with clay know that when soil scientists do something with it they not only clean it, but they also separate it into fractions. Normally, they select their favourite fraction for examination. What we have done in our work is to choose clays consisting predominantly of montmorillonite-smectite type of minerals. One sample was predominantly illite and one was predominantly kaolinite. We have not been too particular on the purity of the clay. On the other hand, we have also realised that fractionation of particle size is a problem. Particle size becomes a problem when the major part of the exchange sites or sorption sites of the clay is on the surface. For clays like bentonite or smectite, for which the high ion exchange capacity is related to a volume effect in the clay, that is to substitution in the lattice proper, we expect that, under normal conditions, the properties will be relatively independent of particle size. But other minerals, for example kaolinite, which has probably most sorption sites on the surface of the particles will show great variations of sorption capacity with particle size.

W. BOCOLA, Italy

I have a question for Mr. Bo, which relates to a comment made a few minutes ago by M. Bonne. Clay is considered as a geological medium for disposal of radioactive waste. If we look at the situation in which glasses are slowly leached by the water present in the clay, not considering the possibility of flooding of the repository, in my opinion, it is possible that the concentrations in the leaching solution will increase to values larger than those normally considered, not only for the radionuclides, but also for other inactive chemicals from the glass matrix or from the container materials. Have you considered this point in regard to the Kd and cation exchange capacities of clays?

P. BO, Denmark

First of all, I think that we have been discussing geological barriers, distinguishing between the immediate neighborhood of the disposal site and the major geological barriers, far away from the repository. Far from the repository, the ground water composition should be close to the normal composition because most of the components released within the clay, that is the silica from the glass, the iron from the stainless steel casing and the various cations from the waste, will have been precipitated, being retained in the primary barriers. This means that at a certain distance from the repository they will be present in very minute quantities.

We have not considered the immediate neighborhood of the repository in connection with ion exchange studies. I believe that processes such as crystallisation, precipitation and so on, are more important in the immediate neighborhood of the repository.

F.P. SARGENT, Canada

Are there any more questions or comments that you would like to address to this issue of the use of clays as backfilling, buffering, and sealing materials?

P. BO, Denmark

I have the feeling that this issue has already been addressed a few times before. For the specific matter of the cation exchange capacity, it is a fact that different measurement methods give different values. Scientists try to show that they get good agreement between different methods. The previous speaker for example could characterise the clays in terms of retention index, and this retention index followed quite nicely the cation exchange capacity, which is what you would expect in these systems; the highest was the bentonite, in the middle was the illite and the lowest was the kaolinite. If you try to make any sort of thermo-dynamic description of the systems, you need the cation exchange capacity, and you should be careful to measure it in a way that is the most realistic.

Of course, all of this is very far from what nature is doing, because the processes you observe in nature are very complex. Very often, you are not able to differentiate between the different components affecting these processes and the different mechanisms which are involved. So, this is the point where fundamental studies come in. They cannot be taken for themselves alone, but in combination with what we see in nature, they will be of help.

L.R. DOLE, United States

Perhaps to dramatise that last point and to get away from ion exchange as a dominant subject of discussion, I would like to make a comment on the last French paper, with regard to the reten-tion of protechtonate and iodide. The work of Francis and Bondietti at Oak Ridge and the work at Battelle Pacific Northwest indicate very strongly that the immobilisation of the protechtonate ion is a redox phenomonon, particularly with some of the basalts of interest to our programme. So here is a clearly non-ion exchange mechanism for immobilisation. Second, my work at Westinghouse, which unfortunately is proprietary, and the work by Freed and Freeman at Argonne show a non-ion exchange interaction with iodine. There are two potential mechanisms: iodine forms specially strong complexes with B metals and it has the capacity to form a donor-acceptor charge compound with sulphur. So, here are two potential mechanisms that make these two materials stand out. So, there are other very important mechanisms for immobilisa-tion.

F.P. SARGENT, Canada

Could I ask you what the B metals are?

L.R. DOLE, United States

I am talking about lead, mercury, silver, copper, etc.

F.P. SARGENT, Canada

We in Canada have done some work along these lines. The only work that has been done in detail was related to iodine on lead oxide-lead carbonate mixtures. The X ray diffraction work suggest that an iodide is formed. As yet, we have no information as to its rate of release, since our experiments are quite short term. Anybody who anticipates doing some of these experiments, must appreciate that the answers will be very dependent on the concentrations of the species being used. For example, we attempted to repeat some work and found that we could not. We could not because we used a 10^{-6} molar solution, and the previous experiment had been carried out with a 10^{-13} molar solution. My point is that you might find that certain lead minerals will absorb iodine, but the overall total capacity might be quite small.

P. COHEN, France

A propos de mobilisation des ions sur les argiles, nous avons fait des expériences analogues à celles que M. Bo nous a décrites, avec les argiles du plateau de Saclay, argiles qui sont constituées par un mélange d'illite, de caolinite et de montmorillonite. Les études qui ont été faites sont des études d'équilibre entre les argiles saturés avec du calcium ou avec un cation monovalent comme l'ammoniaque et en solution un ion monovalent et un ion divalent-strontium et césium. À partir de quantités très faibles jusqu'à des quantités de macro-composants importantes on suivait les équilibres chimiques avec le traceur radioactif. De l'ensemble de ces études, il résulte deux choses: on a constaté que dans ce cas, le strontium suivait parfaitement la loi d'action de masse et que l'équilibre était un vrai équilibre d'échange d'ion, c'est-à-dire que si on fixe un certain nombre de milliéquivalents, on retrouve en solution un nombre équivalent de milliéquivalents de l'autre ion, donc il y a un parfaite cohérence et une analogie totale avec une résine organique échangeuse d'ions par définition. Tandis que, dans le cas du césium, ce n'était pas du tout cette observation que l'on a faite, ce qui montre, ce qui n'a d'ailleurs aucune originalité, que le mode de fixation du césium sur les argiles est beaucoup plus complexe.

P. BO, Denmark

I am glad to hear that the strontium obeys the normal laws. This is probably due to the fact that you have a mixture of montmorillonite, illite and kaolinite in which the ion exchange capacity is dominated by the illite and montmorillonite. The special problems with cesium seem to be like the ones that we had with kaolinite. I would like to emphasise that I showed data on kaolinite. I expect that when we come to a measurement of montmorillonite, we will get somewhat better agreement with ion exchange theory. With illite, I expect that we will get special problems with cesium, because illite has a special affinity for the monovalent ions in which the ammonium and the potassium ions play a special role. Cesium is likely to behave something like that.

D. RAI, United States

I am a soil chemist myself, and I feel the necessity of characterising the soils with the cation exchange capacity. But I feel that it has been blown out of proportion. Most classical studies on ion exchange are done at a relatively high concentration. They follow an ion exchange behaviour. When we talk about trace concentrations, the tracer quantity versus the ion exchange capacity is very small. Then, the ion exchange behaviour may not be followed. Number two, the ion exchange capacity is a variable factor. When you determine the ion exchange capacity at a low and high pH they do not agree. Third, the ion exchange capacity is just a relative number. It is just a means of characterising soils. Number four, it may only work with certain ions, as for example strontium, calcium, magnesium. It has never been shown that it works in the case of trivalent ions. It will probably never be shown that it works in the case of plutonium, americium or some of these other ions.

F.P. SARGENT, Canada

I know that Peter Bo is dying to reply to this, but we have other things to discuss. I see no reason why you should not make a very short reply. Would you like to? However, bear in mind that we have 20 minutes and we have a lot more to do. It seems to me that this is a quarrel between soil scientists that could possibly be solved later.

P. BO, Denmark

My remark is about one of the points that you raised, the problem of the concentration levels. You have to bear in mind that we are trying to study the sorption of radionuclides at very low concentrations. This sorption is a function of the ground water composition, which in general is high in dissolved salts. So, the properties of the clay are determined to some degree by the macro-chemistry, the ground water composition. We aim at determining the sorption capacity for tracer amounts, when the system is fixed by the macro-chemistry.

F.P. SARGENT, Canada

In essence we have about 10 more minutes for technical discussion. I suggest that we discuss during a few minutes the modelling problems raised by several papers.

F. GIRARDI, CEC

We have presented part of these modelling studies. I am strongly convinced that at this stage the results of these modelling exercises cannot be exploited to rule out any validity of any type of barrier. They cannot be used to indicate that any radionuclide is not important; that it is not a dominant source of a hazard. I think that we are still at a very early stage, and experimental results are being obtained which can, for instance, change completely the scale of hazardous nuclides. So, I would strongly recommend that we do not take too seriously the results of the present hazard evaluations. They are simply exercises. They have no realtionship, yet, with nature or with a particular repository.

M.D. HILL, United Kingdom

Well, I would like to comment on that. My comments about neptunium and plutonium were not based on modelling studies or sorption characteristics. They were based on the basic dosimetric data from ICRP, showing that there are two or three orders of magnitude of difference in the dose per unit intake of plutonium and neptunium. On that alone, neptunium is more important.

L.R. DOLE, United States

To follow the philosophy of Girardi's comment, I think in this problem, even the U.S. has finite capacity in money. There have to be priorities. There has to be some traceable way to establish those priorities. The best tool we have is the use of models. Realising their inadequacies, we must be ready to change the direction of the programme, when it becomes necessary. But, you have to use this kind of evaluations to set priorities.

F. GIRARDI, CEC

Well, I am convinced that we all have limited resources and they must be used very carefully. But, for instance, I do not think that we can base any decision on Burkholder's studies. A natural geochemical barrier is necessarily site specific. Burkholder's studies are applicable to the Hanford site, but there are not many places in Europe which are similar to this site. Usually, hazard evaluation studies are quoted to support certain statements, without mentioning that all these studies include always words of caution to the effect that they are simply exercises.

F.P. SARGENT, Canada

I think this meeting is beginning to sound a bit more like a workshop; I regret that it should happen in the last fifteen minutes.

L.R. DOLE, United States

To the point that somewhere engineering judgement is used in setting priorities, the trend is to hide behind the most complex model that you can find. Applying the Burkholder model implies that you have made the judgement that in the cases most important to your programme, the set of conditions to which this model applies are sufficiently close. So, you hide that engineering judgement behind all the calculations and all the pages of computer printout. But essentially, there is a judgement involved.

F.P. SARGENT, Canada

I believe that the German delegation would like to introduce a statement in the proceedings of this Workshop. Afterwards Dr. Gera has somme communications.

G. MATTHESS, F.R. of Germany

It was a good opportunity for us to see the state of the art of investigations on migration of long-lived radionuclides out of repositories.

The topic can be divided into two areas: modelling techniques and the measurements of parameters of the geological medium. Concerning the models, we need a sensitivity analysis for a great variety of parameters to find the critical parameter con-stellation for a fast transport of radionuclides. The number of parameters should not be restricted from the beginning, so that the neglect of a possible important one can be avoided.

Concerning the measurement of model parameters, two main techniques are used, which have different aims. Laboratory experiments with pure geologic materials and well defined solutions provide information on the physical-chemical processes.

These techniques are well developed but the discrepancies of the results show the need for experimental procedures which approach the natural conditions, as much as possible. This should be achieved by intensified efforts in the field and by an inter-comparison of results.

The use of undisturbed geological materials for laboratory experiments may be considered as a helpful link.

In situ studies should be done in representative areas and geological materials so that the results may be of use for the assessment of other sites.

Naturally occurring radionuclides and radionuclides released to the environment by the action of man should be studied also (fallout-studies, mines, accidents, planned releases, etc.)

F. GERA, NEA

I do not have any comments on the present subject of discussion, but I would like to take this opportunity to inform you of two current initiatives by NEA. I think that they might be of interest to this group.

The first one is the Radionuclide Migration Newsletter. This is a follow-up of the meeting that took place in Studsvik, last September. At the time, it was decided to have a Newsletter on Radionuclide Migration in Crystalline Rocks. Afterwards, this was discussed within the Coordinating Group on Geologic Disposal, and it was decided that it might be worthwhile to expand this Newsletter to cover all rock types. According to the timing of this initiative, I should send out the first call for input reports by the middle of March. The Newsletter should be issued some time around May. I think that I will use the list of this meeting's participants to expand the distribution list of the Newsletter. You might be asked by your respective member of the Coordinating Group to provide input data for this Newsletter.

Another item I would like to mention is the idea of setting up an information system in the field of safety assessment and migration studies, related to geologic disposal of radioactive waste. It may be based on a data bank similar to the one being developed in Battelle Pacific Northwest Laboratory. In two week's time in Richland, there will be a small meeting with the purpose of discussing the cost effectiveness and the scope of such a data bank. The Commission of the European Communities is interested in this initiative and is planning to participate in the meeting. Therefore, the data bank would be coordinated with their programmes and their activities in this field.

F.P. SARGENT, Canada

I have a comment to make about the meeting. I feel that this was a good meeting. I learned lots of things.

However, by no stretch of the imagination was this a workshop. I suggest that we really do need a workshop. After the IAEA/NEA meeting in Otaniemi, somebody should seriously consider setting up a workshop. It should have a restricted attendance; that would exclude people like me, for example. The people there should be the people who actually do the work. It should be located where work is being done. I suggest that there are several places. One which comes to mind is ISPRA. I feel that you must have an isolated place, so that people are forced to sit down and talk to one another. The only way that you can make people mix, is to isolate them, at least at lunch-time. So, this is my only criticism of the meeting.

Mr. Orlowski would like to make some closing remarks. Before he does that, I would like to say how much I have apprecaited being in Brussels and the excellent facilities for conferences available here.

S. ORLOWSKI, CEC

Avant de nous séparer, permettez moi d'abord de vous remercier pour les intéressantes contributions que vous avez apportées à cette Réunion tant par vos communications que par vos questions et réponses qui furent exceptionnellement nombreuses. Je voudrais ensuite tirer quelques conclusions.

Je m'appuierai en bonne partie sur les notes de M. Heremans - qui a été obligé de nous quitter tout à l'heure -, puisque je n'ai pu moi-même assister à toutes les séances; je retiendrai également certaines suggestions d'autres participants.

Les travaux présentés ont porté tant sur les études expérimentales de la migration des radionucléides dans la géosphère que sur la modélisation de cette migration.

En ce qui concerne les études expérimentales, une nette tendance à travailler sur des cas concrets se dégage; la concentration des efforts se fait sur les roches salines, les roches dures du type granite et l'argile, roches qui constituent le triptyque des études de stockage géologique de la Communauté européenne. Les travaux se déroulent aussi bien en laboratoire qu'in situ.

Dans le cas des laboratoires, un effort vigoureux se manifeste pour reproduire les conditions régnant dans le sous-sol profond. A cette échelle, il est possible de mieux comprendre les phénomènes et même d'apprécier avec une bonne approximation l'importance des différents paramètres. Bien sûr, des zones obscures subsistent, en particulier en ce qui concerne la formation et l'influences des complexes; des études prometteuses sont en cours.

Dans le cas des expériences in situ, qui apparaissent unanimement reconnues comme nécessaires, on observe également un effort pour cerner au mieux le réel; il faut cependant regretter que, pour des raisons dont les scientifiques ne sont pas responsables, ces expériences soient le plus souvent conduites sur des sites représentatifs certes, mais pas sur les sites réellement envisagés.

Il faut également regretter qu'elles soient jusqu'à présent limitées à des profondeurs bien inférieures à celles envisagées pour les enfouissements du futur.

En ce qui concerne la modélisation de la migration, la majorité des pays ont adapté au mieux et avec une grande ingéniosité, les équations de transfert de masse au contexte géologique spécifique qui est le leur. Les paramètres principaux sont maintenant bien repérés et leur importance démontrée.

L'ensemble de ces question fait déjà l'objet d'échanges d'information permanent et de discussions des résultats, entre laboratoires nationaux et laboratoires de notre Centre Commun de Recherches, dans le cadre du programme de Recherches actuel de la Communauté européenne en matière de gestion et stockage des déchets radioactifs. Il l'est également dans un cadre international plus large, en particulier grâce à l'Agence de l'OCDE pour l'Energie Nucléaire.

La confirmation des données expérimentales et des approches théoriques développées par les différents instituts devient cependant de plus en plus importante. Nous proposerons demain à notre Comité de gestion qu'une intercomparaison des modèles théoriques et des résultats expérimentaux fortement accrue soit mise en oeuvre dans le cadre du deuxième programme de la Communauté Européenne. En particulier, des échantillons de minéraux pourraient circuler entre les différents laboratoires intéressés en vue d'effectuer des mesures de sorption.

Bien entendu la Commission des Communautés Européennes est prête à participer à des échanges de ce genre à un niveau international, plus large, par les voies qui seront jugées les plus efficaces. Je pense par exemple à l'accord que nous avons en vue avec l'Atomic Energy of Canada Research Company et bien évidemment à nos amis de l'Agence pour l'Energie Nucléaire de l'OCDE.

F. GERA, NEA

Thank you Mr. Orlowski for your kind words. On behalf of the Nuclear Energy Agency I would like to thank the Commission of the European Communities for hosting so efficiently the meeting. I would also like to thank all the participants in the discussions, all the speakers and of course the Chairmen and the interpreters.

LIST OF PARTICIPANTS

LISTE DES PARTICIPANTS

AUSTRIA - AUTRICHE

TSCHURLOVITS, M., Dr., Atomic Institute of Austrian Universities,
Schnettelstrasse 115, A-1020 Vienna

BELGIUM - BELGIQUE

BAETSLE, L.H., Centre d'Etude de l'Energie Nucléaire, CEN/SCK,
Boeretang 200, B-2400 Mol

BONNE, A.A., Centre d'Etude de l'Energie Nucléaire, CEN/SCK,
Boeretang 200, B-2400 Mol

DE REGGE, P.P., Centre d'Etude de l'Energie Nucléaire, CEN/SCK,
Boeretang 200, B-2400 Mol

HEREMANS, R.H., Centre d'Etude de l'Energie Nucléaire, CEN/SCK,
Boeretang 200, B-2400 Mol

CANADA

SARGENT, F.P., Dr., Atomic Energy of Canada Limited, Whiteshell
Nuclear Research Establishment, Pinawa, Manitoba ROE 1LO

TREMAINE, P., Dr., Atomic Energy of Canada Limited, Whiteshell
Nuclear Research Establishment, Pinawa, Manitoba ROE 1LO

DENMARK - DANEMARK

BO, P., Risø National Laboratory, Chemistry Department,
DK-4000 Roskilde

NILSSON, K.K., Ms., Risø National Laboratory, Chemistry Department,
DK-4000 Roskilde

FRANCE

BARBREAU, A.F., Commissariat à l'Energie Atomique, Institut de
protection et sûreté nucléaire, CSDR, B.P. n° 6,
92260 Fontenay-aux-Roses

BOULANGER, Géostock, Tour Aurore Cedex 5, 92080 Paris la Défense

BOURDEAU, F., Mme, Electricité de France, Direction de l'équipement,
Département sites-environnement-information, 3 rue de Messine,
75008 Paris

COHEN, P., Commissariat à l'Energie Atomique, Office de gestion
des déchets, 29 rue de la Fédération, 75015 Paris

DESPRES, A., Commissariat à l'Energie Atomique, CPr/SPS, B.P. n° 6,
92260 Fontenay-aux-Roses

GUIZERIX, J., Commissariat à l'Energie Atomique, Centre d'Etudes
   Nucléaires de Grenoble, B.P. n° 85, Centre de Tri,
   38041 Grenoble Cedex

KHALANSKI, Electricité de France, Etudes et recherches, Service
   applications et l'électricité et environnement, 6 quai Watier,
   78400 Chatou

LEDOUX, E., Ecole Supérieure des Mines de Paris, Centre d'informatique géologique, Laboratoire d'hydrologie mathématique,
   35 rue Saint-Honoré, 77305 Fontainebleau

LEVEQUE, P.C., Laboratoire de radiogéologie, Université de
   Bordeaux I, 33405 Tallence Cedex

MARGAT, Bureau de recherches géologiques et minières, Département
   hydrogéologie, Avenue de Concyr, 45018 Orléans Cedex

PEAUDECERF, P., Bureau de recherches géologiques et minières,
   Département hydrogéologie, Avenue de Concyr, 45018 Orléans
   Cedex

PEYRUS, J-Ch., Commissariat à l'Energie Atomique, Département de
   sûreté nucléaire, B.P. n° 6, 92260 Fontenay-aux-Roses

RANÇON, D.L., Commissariat à l'Energie Atomique, Institut de
   protection et sûreté nucléaire, DSN, Centre d'Etudes Nucléaires
   de Cadarache, B.P. n° 1, 13115 Saint-Paul-lez-Durance

ROCHON, J., Bureau de recherches géologiques et minières, Département
   hydrogéologie, Avenue de Concyr, 45018 Orléans Cedex

SAAS, A., Commissariat à l'Energie Atomique, CPr/SRE/LRT, Centre
   d'Etudes Nucléaires de Cadarache, B.P. n° 1, 13115 Saint-Paul-
   lez-Durance

UZZAN, G., Commissariat à l'Energie Atomique, Département de
   protection, B.P. n° 6, 92260 Fontenay-aux-Roses

VAUBOURG, P., Bureau de recherches géologiques et minières,
   Département hydrogéologie, Avenue de Concyr, 45018 Orléans
   Cedex

FEDERAL REPUBLIC OF GERMANY - REPUBLIQUE FEDERALE D'ALLEMAGNE

ALBERTSEN, M., Dr., Projektstudie Entsorgung (PSE) des Bundes-
   ministeriums für Forschung und Technologie, Grödeweg 17a,
   D-2300 Kiel 1

BEHRENS, H.K., Gesellschaft für Strahlen- und Umweltforschung mbH,
   Ingolstädter Landstrasse 1, D-8042 Neuherberg

BRÜHL, H., Prof., Institut für Angewandte Geologie, Freie
   Universität Berlin, Wichernstrasse 16, D-1000 Berlin 33

BÜTOW, E., Dr., Institut für Kerntechnik der Technischen Universität
   Berlin, Marchstrasse 18, D-1000 Berlin 10

FAUBEL, W., Intitut für Kernchemie, Universität Mainz, Postfach 3980,
   D-6500 Mainz

JAKUBICK, A.T., Kernforschungszentrum Karlsruhe GmbH, Postfach 3640,
   D-7500 Karlsruhe 1

JOCKWER, N., Gesellschaft für Strahlen- und Umweltforschung mbH
    München, Institut für Tieflagerung, Wissenschaftliche Abteilung,
    Berlinerstrasse 2, D-3392 Clausthal-Zellerfeld

MATTHESS, G., Prof., Geologisch-Paläontologisch, Institut Universität
    Kiel, Olshausenstrasse 40/60, D-2300 Kiel 1

TIETZE, K., Dr., Bundesanstalt für Geowissenschaften und Rohstoffe,
    Stilleweg 2, D-3000 Hannover

ITALY - ITALIE

BOCOLA, W., Dr., Comitato Nazionale per l'Energia Nucleare, CSN
    Casaccia, Radioactive Waste Laboratory, C.P. n° 2400, 00100 Roma

LENZI, G., Dr., Comitato Nazionale per l'Energia Nucleare, CSN
    Casaccia, Radioactive Waste Laboratory, C.P. n° 2400, 00100 Roma

THE NETHERLANDS - PAYS-BAS

GLASBERGEN, P., National Institute for Water Supply, P.O. Box 150,
    2260 AD Leidschendam

HAMSTRA, J., Netherlands Energy Research Foundation ECN,
    Westerduinweg 3, Petten N.H.

VAN DALEN, A., Netherlands Energy Research Foundation ECN,
    Westerduinweg 3, Petten N.H.

VAN DORP, F., Institute for Atomic Sciences in Agriculture,
    P.O. Box 48, Wageningen

SWEDEN - SUEDE

ANDERSSON, K., Ms., Department of Nuclear Chemistry, Chalmers
    University of Technology, S-412 96 Göteborg

LANDSTRÖM, K.O.L., Studsvik Energiteknik AB, S-611 82 Nyköping

TORSTENFELT, B., Department of Nuclear Chemistry, Chalmers
    University of Technology, S-412 96 Göteborg

SWITZERLAND - SUISSE

BECK, R.H., Dr., National Genossenschaft für die Lagerung
    Radioaktiver Abfälle (NAGRA), Parkstrasse 23, CH-5401 Baden

HADERMANN, J., Swiss Federal Institute for Reactor Research, E.I.R.,
    CH-5303 Würenlingen

THURY, M., Dr., National Genossenschaft für die Lagerung
    Radioaktiver Abfälle (NAGRA), Parkstrasse 23, CH-5401 Baden

UNITED KINGDOM - ROYAUME-UNI

HILL, M.D., Ms., National Radiological Protection Board, Harwell,
    Didcot, Oxon. OR11 ORQ

KINSEY, D., United Kingdom Atomic Energy Authority, AERE, Harwell,
    Oxfordshire OX11 ORA

UNITED STATES - ETATS-UNIS

DOLE, L.R., Dr., Oak Ridge National Laboratory, c/o Gesellschaft
    für Strahlen- und Umweltforschung mbH München, Institut für
    Tieflagerung, Wissenschaftliche Abteilung, Berlinerstrasse 2,
    D-3392 Clausthal-Zellerfeld (Federal Republic of Germany)

RAI, D., Dr., Battelle Pacific Northwest Laboratory, P.O. Box 999,
    Richland, WA. 99352

COMMISSION OF THE EUROPEAN COMMUNITIES
COMMISSION DES COMMUNAUTES EUROPEENNES

d'ALESSANDRO, M, Dr., CEC, Chemical Division, Joint Research Center,
    Ispra (Varese), Italy

AVOGADRO, A., Dr., CEC, Chemical Division, Joint Research Center,
    Ispra (Varese), Italy

BERTOZZI, G., Dr., CEC, Chemical Division, Joint Research Center,
    Ispra (Varese), Italy

DEPLANO, A., CEC, Chemical Division, Joint Research Center,
    Ispra (Varese), Italy

DUTAILLY, L, CEC, DG Employment and Social Affairs, Health and
    Safety Directorate, Luxembourg, Grand Duché

FALKE, W., CEC, Nuclear Fuel Cycle Division, 200 rue de la Loi,
    B-1049 Brussels, Belgium

GIRARDI, F., Dr., CEC, Chemical Division, Joint Research Center,
    Ispra (Varese), Italy

HAIJTINK, B., CEC, Nuclear Fuel Cycle Division, 200 rue de la Loi,
    B-1049 Brussels, Belgium

MYTTENAERE, C., Dr., CEC, DG XII - Biology, Radiation Protection and
    Medical Research, 200 rue de la Loi, B-1049 Brussels, Belgium

ORLOWSKI, S., CEC, Nuclear Fuel Cycle Division, 200 rue de la Loi,
    B-1049 Brussels, Belgium

SALTELLI, A., Dr., CEC, Chemical Division, Joint Research Center,
    Ispra (Varese), Italy

VENET, P., Dr., CEC, Nuclear Fuel Cycle Division, 200 rue de la
    Loi, B-1049 Brussels, Belgium

INTERNATIONAL ATOMIC ENERGY AGENCY
AGENCE INTERNATIONALE DE L'ENERGIE ATOMIQUE

SCHNEIDER, K.J., IAEA, Division of Nuclear Safety and Environmental
    Protection, Kärntnerring 11, A-1011 Vienna, Austria

SECRETARIAT

GERA, F., Dr., OECD Nuclear Energy Agency, Radiation Protection
    and Waste Management Division, 38 boulevard Suchet,
    75016 Paris, France

DELLA LOGGIA, V.E., Dr., CEC, Nuclear Fuel Cycle Division,
    200 rue de la Loi, B-1040 Brussels, Belgium

SOME
NEW PUBLICATIONS
OF NEA

QUELQUES
NOUVELLES PUBLICATIONS
DE L'AEN

ACTIVITY REPORTS

RAPPORTS D'ACTIVITÉ

Activity Reports of the OECD
Nuclear Energy Agency (NEA)

- 6th Activity Report (1977)

- 7th Activity Report (1978)

Rapports d'activité de l'Agence de
l'OCDE pour l'Energie Nucléaire (AEN)

- 6ème Rapport d'Activité (1977)

- 7ème Rapport d'Activité (1978)

Free on request - Gratuits sur demande

Annual Reports of the OECD
HALDEN Reactor Project

- 17th Annual Report (1976)

- 18th Annual Report (1977)

Rapports annuels du Projet OCDE
de réacteur de HALDEN

- 17ème Rapport annuel (1976)

- 18ème Rapport annuel (1977)

Free on request - Gratuits sur demande

. . .

Twentieth Anniversary of the
OECD Nuclear Energy Agency

- Proceedings on the NEA
Symposium on International
Co-operation in the Nuclear
Field : Perspectives and
Prospects

Vingtième Anniversaire de l'Agence
de l'OCDE pour l'Energie Nucléaire

- Compte rendu du Symposium de
l'AEN sur la coopération internationale dans le domaine nucléaire : bilan et perspectives

Free on request - Gratuit sur demande

NEA at a Glance

Coup d'oeil sur l'AEN

Free on request - Gratuit sur demande

SCIENTIFIC AND TECHNICAL PUBLICATIONS

PUBLICATIONS SCIENTIFIQUES ET TECHNIQUES

NUCLEAR FUEL CYCLE

Uranium - Resources, Production
and Demand

LE CYCLE DU COMBUSTIBLE NUCLEAIRE

Uranium - Ressources, Production
et Demande

1977
£ 4.40, $ 9.00, F 36.00

Reprocessing of Spent Nuclear
Fuels in OECD Countries

Retraitement du combustible
nucléaire dans les pays de l'OCDE

1977
£ 2.50, $ 5.00, F 20.00

Nuclear Fuel Cycle Requirements
and supply considerations,
through the long-term

Besoins liés au cycle du combustible
nucléaire et considérations sur
l'approvisionnement à long terme

1978
£ 4.30, $ 8.75, F 35.00

World Uranium Potential -
An International Evaluation

Potentiel mondial en uranium
Une évaluation internationale

1978
£ 7.80, $ 16.00, F 64.00

. . .

RADIATION PROTECTION

Estimated Population Exposure
from Nuclear Power Production
and Other Radiation Sources

RADIOPROTECTION

Estimation de l'exposition de la
population aux rayonnements
résultant de la production
d'énergie nucléaire et provenant
d'autres sources

1976
£ 1.60, $ 3.50, F 14.00

Personal Dosimetry and Area
Monitoring Suitable for Radon
and Daughter Products
(Proceedings of the NEA
Specialist Meeting, Elliot Lake,
Canada)

Dosimétrie individuelle et
surveillance de l'atmosphère en
ce qui concerne le radon et ses
produits de filiation
(Compte rendu d'une réunion de
spécialistes de l'AEN, Elliot Lake,
Canada)

1976
£ 6.80, $ 14.00, F 56.00

Iodine 129   Iode-129
(Proceedings of an NEA Specialist (Compte rendu d'une réunion de
Meeting, Paris)   spécialistes de l'AEN, Paris)

1977
£ 3.40, $ 7.00, F 28.00

Recommendations for Ionization   Recommandations relatives aux
Chamber Smoke Detectors in   détecteurs de fumée à chambre
Implementation of Radiation   d'ionisation en application des
Protection Standards   normes de radioprotection

1977
Free on request - Gratuit sur demande

Management, Stabilisation and   Gestion, stabilisation et incidence
Environmental Impact of Uranium   sur l'environnement des résidus de
Mill Tailings   traitement de l'uranium
(Proceedings of the Albuquerque   (Compte rendu du Séminaire
Seminar, United States)   d'Albuquerque, Etats-Unis)

1978
£ 9.80, $ 20.00, F 80.00

Radon Monitoring   Surveillance du radon
(Proceedings of the NEA   (Compte rendu d'une réunion de
Specialist Meeting, Paris)   spécialistes de l'AEN, Paris)

1978
£ 8.00, $ 16.50, F 66.00

■ ■ ■

RADIOACTIVE WASTE MANAGEMENT   GESTION DES DECHETS RADIOACTIFS

Bituminization of Low and Medium Conditionnement dans le bitume des
Level Radioactive Wastes   déchets radioactifs de faible et
(Proceedings of the Antwerp   de moyenne activités
Seminar)   (Compte rendu du Séminaire d'Anvers)

1976
£ 4.70, $ 10.00, F 42.00

Objectives, Concepts and   Objectifs, concepts et stratégies
Strategies for the Management   en matière de gestion des déchets
of Radioactive Waste Arising   radioactifs résultant des program-
from Nuclear Power Programmes   mes nucléaires de puissance
(Report by an NEA Group of   (Rapport établi par un Groupe
Experts)   d'experts de l'AEN)

1977
£ 8.50, $ 17.50, F 70.00

Treatment, Conditioning and
Storage of Solid Alpha-Bearing
Waste and Cladding Hulls
(Proceedings of the NEA/IAEA
Technical Seminar, Paris)

Traitement, conditionnement et
stockage des déchets solides alpha
et des coques de dégainage
(Compte rendu du Séminaire
technique AEN/AIEA, Paris)

1977
£ 7.30, $ 15.00, F 60.00

Storage of Spent Fuel Elements
(Proceedings of the Madrid
Seminar)

Stockage des éléments combustibles
irradiés
(Compte rendu du Séminaire de
Madrid)

1978
£ 7.30, $ 15.00, F 60.00

In Situ Heating Experiments in
Geological Formations
(Proceedings of the Ludvika
Seminar, Sweden)

Expériences de dégagement de
chaleur in situ dans les formations
géologiques
(Compte rendu du Séminaire de
Ludvika, Suède)

1978
£ 8.00, $ 16.50, F 66.00

Migration of Long-lived
Radionuclides in the Geosphere
(Proceedings of the Brussels
Workshop)

Migration des radionucléides à
vie longue dans la géosphère
(Compte rendu d'une réunion de
travail de Bruxelles)

1979

■ ■ ■

SAFETY

SURETE

Safety of Nuclear Ships
(Proceedings of the Hamburg
Symposium)

Sûreté des navires nucléaires
(Compte rendu du Symposium de
Hambourg)

1978
£ 17.00, $ 35.00, F 140.00

■ ■ ■

SCIENTIFIC INFORMATION

INFORMATION SCIENTIFIQUE

Neutron Physics and Nuclear Data
for Reactors and other Applied
Purposes
(Proceedings of the Harwell
International Conference)

La physique neutronique et les
données nucléaires pour les réacteurs et autres applications
(Compte rendu de la Conférence
Internationale de Harwell)

1978
£ 26.80, $ 55.00, F 220.00

LEGAL PUBLICATIONS

PUBLICATIONS JURIDIQUES

Convention on Third Party Liability in the Field of Nuclear Energy - incorporating provisions of Additional Protocole of January 1964	Convention sur la responsabilité civile dans le domaine de l'énergie nucléaire - Texte incluant les dispositions du Protocole addition- nel de janvier 1964

1960

Free on request - Gratuit sur demande

Nuclear Legislation, Analytical Study : "Nuclear Third Party Liability" (revised version)	Législations nucléaires, étude analytique : "Responsabilité civile nucléaire" (version révisée)

1977

£ 6.00, $ 12.50, F 50.00

Nuclear Law Bulletin (Annual Subscription - two issues and supplements)	Bulletin de Droit Nucléaire (Abonnement annuel - deux numéros et suppléments)

£ 4.40, $ 9.00, F 36.00

Index of the first twenty issues of the Nuclear Law Bulletin	Index des vingt premiers numéros du Bulletin de Droit Nucléaire

Free on request - Gratuit sur demande

Licensing Systems and Inspection of Nuclear Installations in NEA Member Countries (two volumes)	Régime d'autorisation et d'inspec- tion des installations nucléaires dans les pays de l'AEN (deux volumes)

Free on request - Gratuit sur demande

NEA Statute	Statuts AEN

Free on request - Gratuit sur demande

■ ■ ■

OECD SALES AGENTS
DÉPOSITAIRES DES PUBLICATIONS DE L'OCDE

ARGENTINA — ARGENTINE
Carlos Hirsch S.R.L., Florida 165,
BUENOS-AIRES, Tel. 33-1787-2391 Y 30-7122

AUSTRALIA — AUSTRALIE
Australia & New Zealand Book Company Pty Ltd.,
23 Cross Street, (P.O.B. 459)
BROOKVALE NSW 2100   Tel. 938-2244

AUSTRIA — AUTRICHE
Gerold and Co., Graben 31, WIEN I. Tel. 52.22.35

BELGIUM — BELGIQUE
LCLS
44 rue Otlet. B1070 BRUXELLES -Tel. 02-521 28 13

BRAZIL — BRÉSIL
Mestre Jou S.A., Rua Guaipà 518,
Caixa Postal 24090, 05089 SAO PAULO 10. Tel. 261-1920
Rua Senador Dantas 19 s/205-6, RIO DE JANEIRO GB.
Tel. 232-07. 32

CANADA
Renouf Publishing Company Limited,
2182 St. Catherine Street West,
MONTREAL, Quebec H3H 1M7 Tel. (514) 937-3519

DENMARK — DANEMARK
Munksgaards Boghandel,
Nørregade 6, 1165 KØBENHAVN K. Tel. (01) 12 85 70

FINLAND — FINLANDE
Akateeminen Kirjakauppa
Keskuskatu 1, 00100 HELSINKI 10. Tel. 625.901

FRANCE
Bureau des Publications de l'OCDE,
2 rue André-Pascal, 75775 PARIS CEDEX 16. Tel. (1) 524.81.67
Principal correspondant :
13602 AIX-EN-PROVENCE : Librairie de l'Université.
Tel. 26.18.08

GERMANY — ALLEMAGNE
Alexander Horn,
D - 6200 WIESBADEN, Spiegelgasse 9
Tel. (6121) 37-42-12

GREECE — GRÈCE
Librairie Kauffmann, 28 rue du Stade,
ATHÈNES 132. Tel. 322.21.60

HONG-KONG
Government Information Services,
Sales and Publications Office, Beaconsfield House, 1st floor,
Queen's Road, Central.   Tel. H-233191

ICELAND — ISLANDE
Snaebjörn Jònsson and Co., h.f.,
Hafnarstraeti 4 and 9, P.O.B. 1131, REYKJAVIK.
Tel. 13133/14281/11936

INDIA — INDE
Oxford Book and Stationery Co.:
NEW DELHI, Scindia House. Tel. 45896
CALCUTTA, 17 Park Street. Tel. 240832

ITALY — ITALIE
Libreria Commissionaria Sansoni:
Via Lamarmora 45, 50121 FIRENZE. Tel. 579751
Via Bartolini 29, 20155 MILANO. Tel. 365083
Sub-depositari:
Editrice e Libreria Herder,
Piazza Montecitorio 120, 00 186 ROMA. Tel. 674628
Libreria Hoepli, Via Hoepli 5, 20121 MILANO. Tel. 865446
Libreria Lattes, Via Garibaldi 3, 10122 TORINO. Tel. 519274
La diffusione delle edizioni OCSE è inoltre assicurata dalle migliori
librerie nelle città più importanti.

JAPAN — JAPON
OECD Publications and Information Center
Akasaka Park Building, 2-3-4 Akasaka, Minato-ku,
TOKYO 107. Tel. 586-2016

KOREA - CORÉE
Pan Korea Book Corporation,
P.O.Box n° 101 Kwangwhamun, SÉOUL. Tel. 72-7369

LEBANON — LIBAN
Documenta Scientifica/Redico,
Edison Building, Bliss Street, P.O.Box 5641, BEIRUT.
Tel. 354429—344425

MEXICO & CENTRAL AMERICA
Centro de Publicaciones de Organismos Internacionales S.A.,
Alfonso Herrera N° 72, 1er Piso,
Apdo. Postal 42-051, MEXICO 4 D.F.

THE NETHERLANDS — PAYS-BAS
Staatsuitgeverij
Chr. Plantijnstraat
'S-GRAVENHAGE. Tel. 070-814511
Voor bestellingen: Tel. 070-624551

NEW ZEALAND — NOUVELLE-ZÉLANDE
The Publications Manager,
Government Printing Office,
WELLINGTON: Mulgrave Street (Private Bag),
World Trade Centre, Cubacade, Cuba Street,
Rutherford House, Lambton Quay. Tel. 737-320
AUCKLAND: Rutland Street (P.O.Box 5344), Tel. 32.919
CHRISTCHURCH: 130 Oxford Tce (Private Bag), Tel. 50.331
HAMILTON: Barton Street (P.O.Box 857), Tel. 80.103
DUNEDIN: T & G Building, Princes Street (P.O.Box 1104),
Tel. 78.294

NORWAY — NORVÈGE
J.G. TANUM A/S,
P.O. Box 1177 Sentrum, OSLO 1

PAKISTAN
Mirza Book Agency, 65 Shahrah Quaid-E-Azam, LAHORE 3.
Tel. 66839

PORTUGAL
Livraria Portugal, Rua do Carmo 70-74,
1117 LISBOA CODEX.
Tel. 360582/3

SPAIN — ESPAGNE
Mundi-Prensa Libros, S.A.
Castelló 37, Apartado 1223, MADRID-1. Tel. 275.46.55
Libreria Bastinos, Pelayo, 52, BARCELONA 1. Tel. 222.06.00

SWEDEN — SUÈDE
AB CE Fritzes Kungl Hovbokhandel,
Box 16 356, S 103 27 STH, Regeringsgatan 12,
DS STOCKHOLM. Tel. 08/23 89 00

SWITZERLAND — SUISSE
Librairie Payot, 6 rue Grenus, 1211 GENÈVE 11. Tel. 022-31.89.50

TAIWAN — FORMOSE
National Book Company,
84-5 Sing Sung Rd., Sec. 3, TAIPEI 107. Tel. 321.0698

UNITED KINGDOM — ROYAUME-UNI
H.M. Stationery Office, P.O.B. 569,
LONDON SEI 9 NH. Tel. 01-928-6977, Ext. 410 or
49 High Holborn, LONDON WC1V 6 HB (personal callers)
Branches at: EDINBURGH, BIRMINGHAM, BRISTOL,
MANCHESTER, CARDIFF, BELFAST.

UNITED STATES OF AMERICA
OECD Publications and Information Center, Suite 1207,
1750 Pennsylvania Ave., N.W. WASHINGTON. D.C. 20006.
Tel. (202) 724-1857

VENEZUELA
Libreria del Este, Avda. F. Miranda 52, Edificio Galipàn,
CARACAS 106. Tel. 32 23 01/33 26 04/33 24 73

YUGOSLAVIA — YOUGOSLAVIE
Jugoslovenska Knjiga, Terazije 27, P.O.B. 36, BEOGRAD.
Tel. 621-992

Les commandes provenant de pays où l'OCDE n'a pas encore désigné de dépositaire peuvent être adressées à :
OCDE, Bureau des Publications, 2 rue André-Pascal, 75775 PARIS CEDEX 16.
Orders and inquiries from countries where sales agents have not yet been appointed may be sent to:
OECD, Publications Office, 2 rue André-Pascal, 75775 PARIS CEDEX 16.

PUBLICATIONS DE L'OCDE
2, rue André-Pascal, 75775 Paris Cedex 16

N° 41 160 1979

IMPRIMÉ EN FRANCE
(66 79 05 3) ISBN 92-64-01925-1